T0335840

Principles and Applications of Socio-Cognitive and Affective Computing

S. Geetha
Vellore Institute of Technology, Chennai, India

Karthika Renuka
PSG College of Technology, India

Asnath Victy Phamila
Vellore Institute of Technology, Chennai, India

Karthikeyan N.
Syed Ammal Engineering College, India

A volume in the Advances in
Computational Intelligence and
Robotics (ACIR) Book Series

Published in the United States of America by
  IGI Global
  Engineering Science Reference (an imprint of IGI Global)
  701 E. Chocolate Avenue
  Hershey PA, USA 17033
  Tel: 717-533-8845
  Fax: 717-533-8661
  E-mail: cust@igi-global.com
  Web site: http://www.igi-global.com

Copyright © 2023 by IGI Global. All rights reserved. No part of this publication may be
reproduced, stored or distributed in any form or by any means, electronic or mechanical, including
photocopying, without written permission from the publisher.
Product or company names used in this set are for identification purposes only. Inclusion of the
names of the products or companies does not indicate a claim of ownership by IGI Global of the
trademark or registered trademark.

Library of Congress Cataloging-in-Publication Data

Names: Geetha, S., 1979- editor. | Renukay, D. Karthika, 1981- editor. |
  Phamila, Asnath Victy, 1978- editor. | Karthikeyan, N., 1978- editor.
Title: Principles and applications of socio-cognitive and affective
  computing / S. Geetha, D. Karthika Renukay, Asnath Victy Phamila, and N.
  Karthikeyan, editors.
Description: Hershey PA : Engineering Science Reference, [2022] | Includes
  bibliographical references and index. | Summary: "This book offers
  general information in cognitive science, neurology, and psychology, as
  well as active research in affective computing, in the hopes of helping
  readers better comprehend affective computing"-- Provided by publisher.
Identifiers: LCCN 2022007242 (print) | LCCN 2022007243 (ebook) | ISBN
  9781668438435 (h/c) | ISBN 9781668438442 (s/c) | ISBN 9781668438459
  (ebook)
Subjects: LCSH: Soft computing. | Human-computer interaction. | Affect
  (Psychology)--Computer simulation. | Social psychology--Data processing.
  | Affective neuroscience.
Classification: LCC QA76.9.S63 P75 2022 (print) | LCC QA76.9.S63 (ebook)
  | DDC 006.3--dc23/eng/20220422
LC record available at https://lccn.loc.gov/2022007242
LC ebook record available at https://lccn.loc.gov/2022007243

This book is published in the IGI Global book series Advances in Computational Intelligence and
Robotics (ACIR) (ISSN: 2327-0411; eISSN: 2327-042X)

British Cataloguing in Publication Data
A Cataloguing in Publication record for this book is available from the British Library.

All work contributed to this book is new, previously-unpublished material.
The views expressed in this book are those of the authors, but not necessarily of the publisher.

For electronic access to this publication, please contact: eresources@igi-global.com.

Advances in Computational Intelligence and Robotics (ACIR) Book Series

Ivan Giannoccaro
University of Salento, Italy

ISSN:2327-0411
EISSN:2327-042X

MISSION

While intelligence is traditionally a term applied to humans and human cognition, technology has progressed in such a way to allow for the development of intelligent systems able to simulate many human traits. With this new era of simulated and artificial intelligence, much research is needed in order to continue to advance the field and also to evaluate the ethical and societal concerns of the existence of artificial life and machine learning.

The **Advances in Computational Intelligence and Robotics (ACIR) Book Series** encourages scholarly discourse on all topics pertaining to evolutionary computing, artificial life, computational intelligence, machine learning, and robotics. ACIR presents the latest research being conducted on diverse topics in intelligence technologies with the goal of advancing knowledge and applications in this rapidly evolving field.

COVERAGE

- Cyborgs
- Heuristics
- Neural Networks
- Natural Language Processing
- Cognitive Informatics
- Fuzzy Systems
- Agent technologies
- Pattern Recognition
- Intelligent Control
- Computational Intelligence

IGI Global is currently accepting manuscripts for publication within this series. To submit a proposal for a volume in this series, please contact our Acquisition Editors at Acquisitions@igi-global.com or visit: http://www.igi-global.com/publish/.

The Advances in Computational Intelligence and Robotics (ACIR) Book Series (ISSN 2327-0411) is published by IGI Global, 701 E. Chocolate Avenue, Hershey, PA 17033-1240, USA, www.igi-global.com. This series is composed of titles available for purchase individually; each title is edited to be contextually exclusive from any other title within the series. For pricing and ordering information please visit http://www.igi-global.com/book-series/advances-computational-intelligence-robotics/73674. Postmaster: Send all address changes to above address. Copyright © 2023 IGI Global. All rights, including translation in other languages reserved by the publisher. No part of this series may be reproduced or used in any form or by any means – graphics, electronic, or mechanical, including photocopying, recording, taping, or information and retrieval systems – without written permission from the publisher, except for non commercial, educational use, including classroom teaching purposes. The views expressed in this series are those of the authors, but not necessarily of IGI Global.

Titles in this Series

For a list of additional titles in this series, please visit: http://www.igi-global.com/book-series/

Controlling Epidemics With Mathematical and Machine Learning Models

Abraham Varghese (Higher College of Technology, Oman) Eduardo M. Lacap, Jr. (Higher College of Technology, Oman) Ibrahim Sajath (Higher College of Technology, Oman) Kamal Kumar (Higher College of Technology, Oman) and Shajidmon Kolamban (Higher College of Technology, Oman)

Engineering Science Reference • © 2023 • 300pp • H/C (ISBN: 9781799883432) • US $270.00

Handbook of Research on Computer Vision and Image Processing in the Deep Learning Era

A. Srinivasan (SASTRA University (Deemed), India)

Engineering Science Reference • © 2023 • 400pp • H/C (ISBN: 9781799888925) • US $325.00

Multidisciplinary Applications of Deep Learning-Based Artificial Emotional Intelligence

Chiranji Lal Chowdhary (Vellore Institute of Technology, India)

Engineering Science Reference • © 2023 • 296pp • H/C (ISBN: 9781668456736) • US $270.00

Revolutionizing Industrial Automation Through the Convergence of Artificial Intelligence and the Internet of Things

Divya Upadhyay Mishra (ABES Engineering College, Ghaziabad, India) and Shanu Sharma (ABES Engineering College, Ghaziabad, India)

Engineering Science Reference • © 2023 • 279pp • H/C (ISBN: 9781668449912) • US $270.00

Convergence of Big Data Technologies and Computational Intelligent Techniques

Govind P. Gupta (National Institute of Technology, Raipur, India)

Engineering Science Reference • © 2023 • 335pp • H/C (ISBN: 9781668452646) • US $270.00

701 East Chocolate Avenue, Hershey, PA 17033, USA
Tel: 717-533-8845 x100 • Fax: 717-533-8661
E-Mail: cust@igi-global.com • www.igi-global.com

Editorial Advisory Board

Sonali Agarwal, *Indian Institute of Information Technology, Allahabad, India*
Ali Ameen, *Lincoln University, Malaysia*
V. Chandrasekar, *Colorado State University, USA*
Sheik Dawood J., *Kryptos Technologies, Chennai, India*
Thayabaran Kathiresan, *University of Zurich, Switzerland*
Selvakumar Manickam, *Universiti Sains Malaysia, Malaysia*
Barath Narayanan, *University of Dayton, USA*
Santhi Thilagam P., *National Institute of Technology, Karnataka, India*
Sheng-Lung Peng, *National Taipei University of Business, Taiwan*
Jafar Ali Ibrahim S., *QIS College of Engineering and Technology, India*

List of Reviewers

Kanmani A., *Syed Ammal Engineering College, India*
Umamaheswari B., *PSG Institute of Management, India*
Akalya Devi C., *PSG College of Technology, India*
Grace Mary Kanaga E., *Karunya Institute of Technology and Sciences, India*
Ashok Kumar L., *PSG College of Technology, India*
Geetha M. P., *Sri Ramakrishna Institute of Technology, India*
Uma Maheswari N., *PSNA College of Engineering and Technology, India*
Suresh Kumar Nagarajan, *Presidency University, Bangalore, India*
Chitra P., *Sri Eshwar College of Engineering, India*
Supraja P., *SRM Institute of Science and Technology, India*
Anuradha R., *Sri Ramakrishna Engineering College, India*

Vidhyapriya R., *PSG College of Technology, India*
Rajamohana S. P., *Pondicherry University, Karaikal, India*
Sujatha, *PSG Institute of Management, India*
Monorama Swain, *Silicon Institute of Technology, India*
Hemalatha T., *PSNA College of Engineering and Technology, India*
Srivinitha V., *Bannari Amman Institute of Technology, India*

Table of Contents

Section 1
Socio-Cognitive Computing

Chapter 1
> *Hemalatha J. J., Department of CSE, AAA College of Engineering and
> Technology, India*
> *Bala Subramanian Chokkalingam, Kalasalingam Academy of Research
> and Education, India*
> *Vivek V., AAA College of Engineering and Technology, India*
> *Sekar Mohan, AAA College of Engineering and Technology, India*

Chapter 2
> *R. Muthuselvi, Kamaraj College of Engineering and Technology, India*
> *G. Nirmala, Kamaraj College of Engineering and Technology, India*

Chapter 3
> *Jerritta Selvaraj, Vels Institute of Science, Technology, and Advanced
> Studies, India*
> *Arun Sahayadhas, Vels Institute of Science, Technology, and Advanced
> Studies, India*

Section 2
Affective Computing

Detailed Table of Contents

Section 1
Socio-Cognitive Computing

Chapter 1

>   *Hemalatha J. J., Department of CSE, AAA College of Engineering and Technology, India*
>   *Bala Subramanian Chokkalingam, Kalasalingam Academy of Research and Education, India*
>   *Vivek V., AAA College of Engineering and Technology, India*
>   *Sekar Mohan, AAA College of Engineering and Technology, India*

During the present span, researchers are developing artificial intelligence and network-enabled sensor-based algorithms for various human-centric smart systems such as driverless cars, smart healthcare, virtual reality, and e-commerce personalized shopping. In the course of the last year, software entity hawkers are wrapping AI into business classes. This chapter is organized into three phases. In the first phase, the authors provide a deep view on architecture of cognitive computing. Then they provide some applications of Cognitive AI. Basically, knowledge-based suits are emerging over a couple of diverse functional usage scenarios: 1) computerizing everyday easy tasks, which consume time and 2) offering significant data to the user application. The application instances can pioneer possibilities to expand the yield and decision-making accuracy. In the third phase, the authors provide some fruitful ongoing research in cognitive computing.

Chapter 2

    R. Muthuselvi, Kamaraj College of Engineering and Technology, India
    G. Nirmala, Kamaraj College of Engineering and Technology, India

Learning aids in the development of attitude. It encourages the individual to learn new skills. It is critical to master three learning domains: cognitive, affective, and psychomotor. The cognitive computing system instantaneously processes data and gives solutions to questions. Affective computing is the development of tools that can recognise, understand, examine, and replicate human brains. Communication and behaviour are impacted by autism spectrum disorder (ASD), a developmental disease. ASD leads to have difficulty in interacting with society and communicating with society. It states that people with ASD have 1) difficulty in conversation and contact with other people, 2) symptoms that interfere with the person's ability to function normally in the society, and 3) restricted interests and repetitive behaviours. In the chapter, a computer-based model is developed for various emotions, facial expressions, and voice and body language. The aim is to develop a computer-based model that supports the autistic children to understand emotions and express their feelings.

Chapter 3

    Jerritta Selvaraj, Vels Institute of Science, Technology, and Advanced
        Studies, India
    Arun Sahayadhas, Vels Institute of Science, Technology, and Advanced
        Studies, India

Computer-based learning and training has increased over the past few decades and has become the norm of today. This allows students to learn from the comfort of their home at their convenient time and pace. The learning materials can be accessed at any time and from anywhere making it easier for students. Recent studies indicate that negative emotions such as anger, frustration, confusion, and boredom inhibit the learning process and positive emotions such as excitement and enthusiasm support the learning process. Many students complain of boredom and similar negative emotions, which deteriorate the rate of learning. This research tries to integrate the emotion recognition system to help students during the learning process. Appropriate coping mechanisms can be integrated into the e-learning system to keep the students engaged and alert in the process. Emotions can be identified using wearable sensors and appropriate learning activity or breaks or physical activity can be given to students based on the emotion experienced by the students.

Chapter 4

Vidhya R., SRM Institute of Science and Technology, India
Sandhia G. K., SRM Institute of Science and Technology, India
Jansi K. R., SRM Institute of Science and Technology, India
Nagadevi S., SRM Institute of Science and Technology, India
Jeya R., SRM Institute of Science and Technology, India

Social, psychological, and emotional well-being are all aspects of mental health. Mental illness can cause problems in daily life, physical health, and interpersonal connections. Severe changes in education, attitude, or emotional management of students cause suffering are defined as children's mental disorders. Artificial intelligence (AI) technology has lately been advanced to help intellectual fitness professionals, especially psychiatrists and clinicians, in making choices primarily based totally on affected person records along with medical history, behavioural records, social media use, and so on. There is a pressing need to address core mental health concerns in children, which can progress to more serious problems if not addressed early. As a result, a shallow learning technique-assisted integrated prediction model (SLIPM) has been presented in this research to predict and diagnose mental illness in children early. Convolutional neural networks (CNN) are built first in the proposed model to learn deep-learned patient behavioural data characteristics.

Chapter 5

Shanthalakshmi Revathy J., Velammal College of Engineering and
Technology, India
Uma Maheswari N., PSNA College of Engineering and Technology, India
Sasikala S., Velammal College of Engineering and Technology, India

Emotion recognition based on biological signals from the brain necessitates sophisticated signal processing and feature extraction techniques. The major purpose of this research is to use the enhanced BiLSTM (E-BiLSTM) approach to improve the effectiveness of emotion identification utilizing brain signals. The approach detects brain activity that has distinct characteristics that vary from person to person. This experiment uses an emotional EEG dataset that is publicly available on Kaggle. The data was collected using an EEG headband with four sensors (AF7, AF8, TP9, TP10), and three possible states were identified, including neutral, positive, and negative, based on cognitive behavioral studies. A big dataset is generated using statistical brainwave extraction of alpha, beta, theta, delta, and gamma, which is then scaled down to smaller datasets using the PCA feature selection technique. Overall accuracy was around 98.12%, which is higher than the present state of the art.

Chapter 6

*Uma N. Dulhare, Muffakham Jah College of Engineering and
     Technology, India*
Shaik Rasool, Methodist College of Engineering and Technology, India

The COVID-19 pandemic raised the need for harnessing digital infrastructure for many healthcare services like appointment scheduling, surveillance, and checking the patients remotely. A digital platform is needed that should be reliable for disease identification and monitoring using IoT, which can compensate for vital activities like the slow rate of viral tests and vaccine development also recognized as apt technology for bridging various devices. Although the technology has been used to connect the daily activities with the physical metrics, forecasting of COVID-19 is very vital and necessary. The fitness measures like body temperature, heartbeat rate, SPo2 from wearable devices can be used to alert the users. Groups of affected individuals can be remotely checked, and data can be collected to analyse the rate of transmission and symptoms. This chapter emphasizes harnessing the potential of IoT by comprehending the importance of IoT in various domains and its various applications. It will explore various IoT devices and focus on challenges and advantages and security aspects.

Section 2
Affective Computing

Chapter 7

*R. Nareshkumar, Department of Networking and Communication, SRM
     Institute of Science and Technology, India*
*G. Suseela, Department of Networking and Communication, SRM
     Institute of Science and Technology, India*
*K. Nimala, Department of Networking and Communication, SRM
     Institute of Science and Technology, India*
*G. Niranjana, Department of Computing Technologies, SRM Institute of
     Science and Technology, India*

The development of the automobile industry and civilian infrastructures improved the lifestyles of everyone in the world. In parallel to the rise in quality of life of everyone, the number of road accidents also rose. The major reason behind road accidents is emotional factors of the drivers. The emotional imbalance will influence the drivers to abandon the traffic rules, neglect speed limits, cross the signals, cross the lane, etc. Recently automobile industries have extended their researches to the development of emotion sensing systems and embedding them inside the vehicles using affective computing technology to mitigate the road accidents. These emotion sensing systems will be decisive and act as human-like driver-assistive systems in alarming the drivers. This chapter focuses on bringing out the feasibility and existing challenges of affective computing in sensing the emotional factors of drivers for improved road safety.

The objective of this work is to evaluate the impact of yoga and meditation intervention on engineering students' stress perception, anxiety levels, and mindfulness skills. Adolescents are experiencing greater stress associated with academic performance, extracurricular activities, and worry about the future. Meditation is the practice by which there is constant observation of the mind. It requires you to focus your mind at one point and make your mind still in order to perceive the 'self'. In this chapter, the authors interrogate the impact of different yogasanas and meditation in enabling learners to get rid of mindful stress. The remainder of the chapter is organized as follows: Section 2 explains the causes of stress and its effects. Section 3 presents the different categories of emotion in stress for explaining the several of levels present in it. Section 4 explains yoga and emotional stress reduction, and Section 5 discusses the conclusions.

Behavioral issues are categorized by means of persistent difficulties faced from the beginning of childhood. The children with this behavioral disorder restrict social communication and show repetitive interest. Some children have challenging behaviors that are beyond their age and identification becomes difficult. These problems can cause temporary stress on a child's health. A lot of children are impacted by behavioral-related issues from birth, and unfortunately, no scientifically backed early detection mechanism is available to identify the stated issues within the first three years. Using AI and data analytics algorithm, the behavioral profile of a child can be analyzed using a key marker to identify behavioral issues later. Cloud-based AI solutions could be used to implement early detection. Machine learning algorithms using SVM have the potential to create the decision boundary for the segregation using possible classes.

A brain-computer interface (BCI) is a computer-based system that collects, analyses, and converts brain signals into commands that are sent to an output device to perform a desired action. BCI is used as an assistive and adaptive technology to track the brain activity. A silent speech interface (SSI) is a system that enables speech communication when an acoustic signal is unavailable. An SSI creates a digital representation of speech by collecting sensor data from the human articulatory, their neural pathways, or the brain. The data from a single stage is very minimal in order to capture for further processing. Therefore, multiple modalities could be used; a more complete representation of the speech production model could be developed. The goal is to detect speech tokens from speech imagery and create a language model. The proposal consists of multiple modalities by taking inputs from various biosignal sensors. The main objective of the proposal is to develop a BCI-based end-to-end continuous speech recognition system.

Many people worldwide have the problem of visual impairment. The authors' idea is to design a novel image captioning model for assisting the blind people by using deep learning-based architecture. Automatic understanding of the image and providing description of that image involves tasks from two complex fields: computer vision and natural language processing. The first task is to correctly identify objects along with their attributes present in the given image, and the next is to connect all the identified objects along with actions and generating the statements, which should be syntactically correct. From the real-time video, the features are extracted using a convolutional neural network (CNN), and the feature vectors are given as input to long short-term memory (LSTM) network to generate the appropriate captions in a natural language (English). The captions can then be converted into audio files, which the visually impaired people can listen. The model is tested on the two standardized image captioning datasets Flickr 8K and MSCOCO and evaluated using BLEU score.

Chapter 12
    Arjun Sharma, Vellore Institute of Technology, Chennai, India
    Hemanth Harikrishnan, Vellore Institute of Technology, Chennai, India
    Sathiya Narayanan Sekar, Vellore Institute of Technology, Chennai,
        India
    Om Prakash Swain, Vellore Institute of Technology, Chennai, India
    Utkarsh Utkarsh, Vellore Institute of Technology, Chennai, India
    Akshay Giridhar, Vellore Institute of Technology, Chennai, India

The initial outbreak of the coronavirus was met with lockdowns being enforced all over the world in March 2020. A prominent change in human lifestyle is the shift of professional and academic work to online platforms, as opposed to previously attending to them in person. As with any major change, the implementation of complete remote work and study is expected to affect different people differently. Through the results of a questionnaire designed as per the implications of the self-efficacy theory shared with people who were either students, working professionals, entrepreneurs, or homemakers aged between 12 and 60 years, the authors perform statistical analysis and subsequently hypothesize how different aspects of remote work affect the population from a mental standpoint using t-test, with respect to their professional or academic work. This is followed by predictive modelling through machine learning algorithms to classify working preference as 'remote' or 'in-person'.

Foreword

This book is aimed to enable Socio Cognitive and Affective Computing, which has been an intensive research area for decades. Many core technologies, such as machine learning, statistical based models, deep learning and affective computing have been developed along the way, mostly prior to the new millennium. These techniques greatly advance the state of the art in Socio Cognitive and Affective Computing and in its related fields. The Socio-Cognitive and Affective Computing is a wide-ranging approach to comprehending human intelligence and machine cognition. It's a research field which aims to investigate research problems and uses of affective computing and socio-cognitive behaviors. Human capabilities can be developed by using these ways to improve their powers of observation, analysis, decision-making, processing, and responding to others and to regular or difficult circumstances. In the recent years, however, we have observed a new surge of interest in Socio Cognitive and Affective Computing. In our opinion, this change was led by the increased demands on Cognitive Computing tools and the success of new applications such as IBM Watson, artificial intelligence tools like expert systems, and intelligent personal assistant tools like Amazon Echo, Apple Siri, Google Assistant, and Microsoft Cortana. Our unique aspect of this book is to broaden the view of bio inspired cognitive computation models. All in all, this book is likely to become a definite reference for researchers and practitioners in the deep learning era. This book masterfully covers the basic concepts required to understand affective computing as a whole, and also in detail the most cutting-edge research happening in this field. The reader of this book will become articulate in the new state of the art of socio cognitive and affective computing technologies.

Regards

Sheng-Lung Peng
College of Innovative Design and Management, National Taipei University of
Business, Taiwan

Foreword

The objective of the book is to provide the readers with the fundamentals, recent trends of socio cognitive and affective computing. This book gives the introduction, applications of socio cognitive and affective computing to the academicians, researchers and students who are new to this field. This book produces evident research outcomes using cutting edge technologies in this field with real time applications. The book provides a refreshing and motivating new synthesis of the field by one of AI's master expositors and leading researchers. The Socio-Cognitive and Affective Computing is a broad approach to comprehending human intelligence and machine cognition. It's a research field which aims to investigate research problems and uses of affective computing and socio-cognitive behaviors. Human capacities can be improved by using these ways to improve their powers of observation, analysis, decision-making, processing, and responding to others and to regular or difficult circumstances. Cognitive computing tools like IBM Watson, artificial intelligence tools like expert systems, and intelligent personal assistant tools like Amazon Echo, Apple Siri, Google Assistant, and Microsoft Cortana can all be used to improve humans' understanding, decision-making, acting, learning, and avoiding problems. Designing this kind of systems requires combining knowledge and methods of ubiquitous and pervasive computing, as well as physiological data measurement and processing, with those of socio-cognitive and affective computing. It explores all aspects of cognitive agents, including perception, action, affective and cognitive learning and memory, attention, decision making and control, social cognition, language processing and communication, reasoning, problem solving, and consciousness.

Best Wishes

L. Ashok Kumar
PSG College of Technology, India

Preface

Emotions are a natural part of our cognitive activity and play a vital role in human interaction and decision-making processes. Emotion is a sequence of events linked together by feedback effects. Feelings and behaviour can have an impact on cognition, and cognition can have an impact on feelings. Emotion can be seen as part of a structural model linked to adaptation. Identifying and analyzing emotional data is essential in a variety of computer science fields, including human-computer interaction, e-learning, medicine, automotive, cyber security, user profiling, and customization. The study of how human process, store, and use information about other people and social environments is known as social cognition. It highlights the importance of mental abilities in our social communication. Cognitive computing, on the other hand, is a term that applies to a model that replicates the working of the human brain and aids in human decision-making. In this perspective, it is a sort of computing aimed at better understanding how the human brain/mind perceives, reasons about, and responds to inputs. As a result, Socio-Cognitive Computing should be viewed as a collection of theoretical multidisciplinary frameworks, techniques, procedures, and hardware/software tools for simulating how the human brain handles social interactions. Affective Computing, a core part of socio-cognitive neuroscience, is the research and development of systems and devices that can detect, understand, analyze, and imitate human affects. Also, it is a multidisciplinary area including computer science, electrical engineering, psychology, and cognitive science. Physiological Computing is a technique that utilizes electrophysiological information acquired from human activity to interface with a computing device. This technology is even more significant When computing becomes pervasively incorporated into everyday life surroundings. As a nutshell, Socio-Cognitive and Affective Computing systems ought to be capable of adjusting their function to the Physiological Computing technique.

Computational Intelligence for Affective and Socio-Cognitive Computing aims at integrating these various albeit complementary fields. Designing this kind of systems requires combining knowledge and methods of ubiquitous and pervasive computing, as well as physiological data measurement and processing, with those of socio-cognitive and affective computing. Because interactions can occur between people and computing agents, an interdisciplinary approach that examines the foundations of affective communication across several platforms is required. A collection of study combining decades of evidence on emotions in psychology, cognition, and neuroscience will inspire creative future research projects and assist in the development of this new subject in the field of affective computing. This reference, which covers several aspects of affective interactions and concepts in affective computing, discusses the fundamentals of emotions, advances our knowledge of the processes of affect in our lives, and suggests by exposing emerging research and exciting techniques for bridging the emotional disparity between humans and machines, all within the context of interactions.

The Science of Intelligence is a broad approach to comprehending human intelligence and machine cognition. The Science of Intelligence transdisciplinary research field which aims to investigate research problems and uses of cognitive computing and socio-cognitive behaviours. The goal is to use machine learning, artificial intelligence, and deep learning frameworks to create data-driven models that can be used to better understand the following objectives:

- Emotional intelligence and mental addiction and the possibility of technological intervention
- Multimodal perception
- Neural plasticity and cognitive compensation
- Implication in developmental and geriatric disorders, and
- Biomarker discovery for cognitive disorders.

This book contains cutting-edge papers detailing groundbreaking fundamental and application research involving bio-inspired computational models of all facets of cognitive systems, as well as affective computing. It opens a unique forum for the sharing of research, current techniques, and global direction in cognitive computation, an evolving topic that spans the gap between disciplines such as sociology, technology, and humanities. This book offers a transdisciplinary platform for the dissemination of cutting-edge research, recent techniques, and future trends in this developing field that combines neuroscience, cognitive psychology, and artificial intelligence. It explores all aspects of cognitive agents, including perception, action, affective and cognitive learning and memory, attention, decision making and control, social cognition, language processing and communication, reasoning, problem solving, and consciousness. These innovations utilize cognitive computing to assist in achieving these desired outcomes:

- Assist people in making better decisions, acting more quickly, and achieving better results.
- Convey pertinent advice and guidance when it's needed.
- Allow people to be more productive and efficient in their actions.
- Enhance wellbeing through reducing error, decreasing loss and damage.
- In complex scenarios marked by uncertainty and risk, cognitive computing can emulate human mental processes and mimic the way the human brain operates.
- Artificial intelligence can accomplish tasks similar to human learning and decision-making. Intelligent personal assistants are capable of recognising voice commands and queries, providing information, and performing specified tasks quickly, efficiently, and effectively.

Human capacities can be improved by using these ways to improve their powers of observation, analysis, decision-making, processing, and responding to others and to regular or difficult circumstances. Cognitive computing tools like IBM Watson, artificial intelligence tools like expert systems, and intelligent personal assistant tools like Amazon Echo, Apple Siri, Google Assistant, and Microsoft Cortana can all be used to improve humans' understanding, decision-making, acting, learning, and avoiding problems. For years, computer science researchers have focused on rational reasoning and ignored the impact of emotions on machine learning, artificial intelligence, and decision making. The only real attempt to create human-like beings has been to build agents with rational decision-making — independent of emotions. As a result of these new findings, end-user emotional experience is becoming a major concern, particularly in human-computer interaction. The ability to create, activate, maintain, and recognize emotions at the end-user level, as well as the ability to imitate emotions, appear to be hot subjects in the near future.

We humbly sought to collect together (so-called unique) types of field information in this book to help with the development of emotional computing with an emphasis on user interactions. We compiled general information in cognitive science, neurology, and psychology, as well as active research in affective computing, in the hopes of helping readers better comprehend affective computing.

The proposed book provides a state-of-the-art idea of the problems and solution guidelines emerging in socio cognitive computing, thus summarizing the roadmap of current machine computational intelligence techniques for affective computing. A wide variety of topics of interest are addressed, from the technical views such as advanced techniques for cognitive computing developments in computer and communication-link environments, in e-learning, emotional data analysis, ambient assisted living etc. We have assembled insights from a representative sample of academicians and practitioners.

The topics in this book are categorized into two sections. The first section provides an insight into the Socio-Cognitive Computing, especially to the way in which it can be employed to deal with autism children, emotion analysis, speech recognition, voice-based captioning and to track COVID with cognitive assistance using IoT devices. The second section presents the affective computing methods usage in behavioural analysis, emotional analysis from EEG signals, predictive modeling of reactions to COVID 19 induced remote work. Each section provides the current research trends in the concerned field of study.

Section 1: Socio Cognitive Computing

Socio Cognitive Computing is a form of computation used to produce more precise theories on how the human brain and mind perceive, reason about, and react to stimulus. Cognitive computing concentrates on impersonating human performance and reckoning to provide solutions to complex problems. Chapter 1 provides the in depth view on architecture of cognitive computing in various technologies such as Internet of Things, Deep Learning and Cloud Computing. It also presents few prominent Cognitive-Artificial Intelligence based applications and insights of fruitful ongoing research in Cognitive Computing.

It is critical to master three learning domains: cognitive (thinking), affective (emotions or feelings), and psychomotor (Physical or kinesthetic). The cognitive dimension is the domain used to evaluate learning and it focuses on knowledge acquisition, retention, and implementation. Chapter 2 focuses on Cognitive model developed for various emotions, facial expressions, voice, and body language. The computer-based model can be repeatedly used to teach the autistic children which will assist the autistic children in understanding the emotions. Chapter 3 presents a methodology to integrate the emotions detected from the Electrocardiogram (ECG) and Electromyograph (EMG) signals into an e-Learning module that would suggest the learner to perform an appropriate activity to normalize an emotion. The various steps in the detection of emotion ECG and EMG are discussed in detail and the GUI for e-Learning using the emotions recognized using ECG and EMG are also presented.

Social, psychological, and emotional well-being are all aspects of mental health. Mental illness can cause problems in daily life, physical health, and interpersonal connections. There is a pressing need to address core mental health concerns in children, which can progress to more serious problems if not addressed early. As a result, a Shallow Learning Techniques Assisted Integrated Prediction Model (SLIPM) has been presented in Chapter 4 to predict and diagnose mental illness in children early. Emotion recognition based on biological signals from the brain necessitates sophisticated signal processing and feature extraction techniques. Chapter 5 has elucidated the use of Enhanced BiLSTM (E-BiLSTM) approach to improve the effectiveness of emotion identification utilizing brain signals. The approach detects brain activity that has distinct characteristics that vary from person to person with great accuracy.

The COVID-19 pandemic raised the need for harnessing digital infrastructure for many healthcare services like appointment scheduling, surveillance, checking the patients remotely, etc. Chapter 6 emphasizes on harnessing the potential of IoT by comprehending the importance of IoT in various domains and its various applications, will explore various IoT devices, focus on challenges and advantages, security aspects, etc. This chapter also discusses the merits of using IoT Technology for Cognitive Assistance in detail.

Section 2: Affective Computing

The study of and creation of technologies for simulating, perceiving, analyzing, and interpreting human affects is known as affective computing and it is a multidisciplinary field that integrates cognitive science, psychology, and computer science. Recently automobile industries have extended their researches to the development of emotion sensing systems and embedding them inside the vehicles using affective computing technology to mitigate the road accidents. These emotion sensing systems will be decisive and act as human like driver assistive system in alarming the drivers. Chapter 7 explores feasibility and existing challenges of affective computing in sensing the emotional factors of drivers for improved road safety. Adolescents are experiencing greater stress associated with academic performance, extracurricular activities and worry about the future. The objective of Chapter 8 is evaluating the impact of yoga and meditation intervention on Engineering students' stress perception, anxiety levels, and mindfulness skills. A lot of children are impacted by behavioral-related issues at the start of birth and unfortunately no scientifically backed early detection mechanism is available to identify the stated issues within first three years. Chapter 9 emphasizes the fact that Using AI and data analytics algorithm, the behavioral profile of a child can be analyzed using a key marker to identify behavioral issues later. It suggests a Cloud based AI solution to implement early detection and assures that Machine learning algorithm using SVM has huge potential to create the decision boundary for the segregation using possible classes.

Brain-Computer Interface (BCI) is a computer-based system that is used as an assistive and adaptive technology to track the brain activity and Silent Speech Interfaces (SSIs) are a sort of assistive technology that helps people regain their ability to communicate verbally. Automatic Speech Recognition (ASR) and related sciences have been hot topics in recent years, and they offer the basis for the Brain Computing Interface. Chapter 10 proposes a BCI based end-to-end continuous speech recognition system based on multimodal fusion framework.

Many people worldwide have the problem of visual impairment. Chapter 11 proposes an assistive technology based on image captioning model for assisting the blind people by using deep learning architecture. From the real-time video the features are extracted using a Convolutional Neural Network (CNN) and the feature vectors are given as input to Long Short-Term Memory (LSTM) network to generate the appropriate captions in a natural language (English). The captions can then be converted into audio file which the visually impaired people can listen. Self-efficacy theory claims that the behavior, environment, and cognitive factors of an individual share high levels of interrelation and Chapter 12 ascertains different influences on one's mindset in a remote setting using statistical analysis techniques and predictive modeling.

Our intention in editing this book is to offer concepts and various techniques that are employed in realizing socio cognitive and affective computing. In editing the book, our attempt is to provide frontier advancements in emotion analysis, application of machine learning to cognitive intelligence implementation using IoT, etc. This book will comprise the latest research from prominent researchers working in this domain. Since the book covers case study-based research findings, it can be quite relevant to researchers, university academics, computing professionals, and probing university students. In addition, it will help those researchers who have interest in this field to keep insight into different concepts and their importance for applications in real life. This has been done to make the edited book more flexible and to stimulate further interest in topics

The topics presented are recent works and research findings in state-of-the-art idea of the problems and solution guidelines emerging in cognitive and affective computing with respect to artificial intelligence perspective, trends and methods, theoretical and mathematical foundations.

The prospective audience for this book will be academicians, researchers, advanced-level students, technology developers, and global consortiums for cognitive and affective computing leading to development of ambient assisted living. This will also be useful to a wider audience of readers in furthering their research exposure to pertinent topics in advancing their own research efforts in this field.

S. Geetha
Vellore Institute of Technology, Chennai, India

Karthika Renuka
PSG College of Technology, India

Asnath Victy Phamila
Vellore Institute of Technology, Chennai, India

Karthikeyan N.
Syed Ammal Engineering College, India

Acknowledgment

Taking up a book project is harder than we thought and more rewarding than we could have ever imagined. None of this would have been possible without the support of our family members, friends and academic connections.

We would like to acknowledge the help of all the people involved in this project and, more specifically, to the authors and reviewers who took part in the review process. Without their support, this book would not have become a reality.

First, we would like to thank each one of the authors for their contributions. Our sincere gratitude goes to the chapter's authors who contributed their time and expertise to this book. We thank all the authors of the chapters for their commitment to this endeavor and their timely response to our incessant requests for revisions.

Second, the editors wish to acknowledge the valuable contributions of the reviewers regarding the improvement of quality, coherence, and content presentation of chapters. Most of the authors also served as referees; we highly appreciate their double task.

The editors would like to recognize the contributions of editorial board in shaping the nature of the chapters in this book. In addition, we wish to thank the editorial staff at IGI Global for their professional assistance and patience.

Thanks to the support from our Management of VIT, PSG College of Technology and Syed Ammal Engineering College and our VC, Pro VC, Dean – SCOPE, colleagues, Principal and peers, without which this book would not exist. Sincere thanks to each one of them.

We want to thank God, Almighty! most of all, because without God we wouldn't be able to do this.

Section 1
Socio–Cognitive Computing

Chapter 1
Cognitive Computing:
A Deep View in Architecture, Technologies, and Artificial Intelligence–Based Applications

Hemalatha J. J.
iD https://orcid.org/0000-0003-0793-4191
Department of CSE, AAA College of Engineering and Technology, India

Bala Subramanian Chokkalingam
Kalasalingam Academy of Research and Education, India

Vivek V.
AAA College of Engineering and Technology, India

Sekar Mohan
AAA College of Engineering and Technology, India

ABSTRACT

During the present span, researchers are developing artificial intelligence and network-enabled sensor-based algorithms for various human-centric smart systems such as driverless cars, smart healthcare, virtual reality, and e-commerce personalized shopping. In the course of the last year, software entity hawkers are wrapping AI into business classes. This chapter is organized into three phases. In the first phase, the authors provide a deep view on architecture of cognitive computing. Then they provide some applications of Cognitive AI. Basically, knowledge-based suits are emerging over a couple of diverse functional usage scenarios: 1) computerizing everyday easy tasks, which consume time and 2) offering significant data to the user application. The application instances can pioneer possibilities to expand the yield and decision-making accuracy. In the third phase, the authors provide some fruitful ongoing research in cognitive computing.

DOI: 10.4018/978-1-6684-3843-5.ch001

Copyright © 2023, IGI Global. Copying or distributing in print or electronic forms without written permission of IGI Global is prohibited.

INTRODUCTION

Cognitive computing is a self-learning scheme that utilizes Machine Learning, Deep Learning, Data Mining, Neural Networks, and Visual Recognition to carry out exactly human-like tasks cleverly and accurately. Cognitive computing concentrates on impersonating human performance and reckoning to provide solutions to complex problems. It gains knowledge with humans more naturally. Cognitive Computing methods classically fall on Deep Learning methods (Min Chen et al., 2018) and Neural Networks. Massive attractiveness of cognitive computing together in intellectual and corporate could leads to fast improvement of software, hardware and artificial intelligence. Cognitive computing logical methods are used in science of behavior, environmental science; signal processing, arithmetic in the direction of building technologies which will be more realistic ability other than the human intelligence. The Powerful and most successful cognitive computing architecture could be employed in assist of 5G network, robotics, deep learning, machine learning, machine architecture, cloud computing and IOT setups. Pattern recognition, machine vision, sustainable development, smart applications and Internet of Things are the few applications of Cognitive computing. Cognitive computing skeleton attached with big data analytics since it hardly needs massive quantity of facts to incorporate dangerous thinking ability of human. Machine learning concepts are highly used in cognitive computing architecture. In addition, in order to construct a successful cognitive computing architecture, techniques such as deep network and machine learning are widely used. The design will describe source, model and computational progression that can normalize the outcome. This may helpful in steganography, augmentation reliability.

Cognitive computing's main persistence is construction of figuring technique which may resolve most tedious complications in short of unvarying humanoid interference. Sequentially, in the direction of carry out this technique in marketable and its extensions, this technique has strongly suggested the subsequent in any sort of systems –

i) Adaptive feature

Adaptive feature is the first most steps in building ML grounded cognitive scheme. Outcome ought to be impersonating human brains capability, characteristic in the direction of training and become accustomed from the environment. Further the systems cannot be individually worked to the inaccessible task. Instead the aforementioned desires in more active in nature for connection, known about the foremost objectives, and necessities.

ii) Inter-active feature

Interactive, is a perceptive outcome is related to the human brain how the human thinks must interact with all the assembled systems like processor, devices, cloud services and user. While talk about the communication cognitive system acts and communicates as bi-directionally. It ought to fully know about the human input and deliver applicable outcomes by means of NLP and other learning strategies. Many accomplished intellectual such as Mitsuku which is highly admired to this feature.

iii) Stateful feature

Iterative and stateful – Is a system should "recall" earlier communications in a progression and yield evidence that is appropriate for the precise application at that fact in time. It must be able to explain the problem by examining questions or discovering an additional source. This feature desires a cautious application of the data quality and authentication practices in demand to guarantee that the system is constantly provided with sufficient information and that the documents sources it activates on to distribute consistent and up-to-date input.

iv) Contextual

Contextual, it essentially recognizes, classify, and extract contextual elements such as meaning, syntax, time, location, appropriate domain, regulations, user's profile, process, task, and goal. It might appeal on multiple causes of evidence, counting both structured and unstructured digital information, as well as sensory inputs (visual, gestural, auditory, or sensor-provided).

COGNITIVE COMPUTING ARCHITECTURE

The system architecture of cognitive computing is shown in Figure 1 and explained below. With the tremendous help of upcoming booming technology such as 5G network (M.Chen et al., 2017), Reinforcement learning, machine learning, deep learning the applications such as facial recognition, human computer interaction, voice recognition, gesture recognition, will be implemented in a huge manner. In addition, environ cognitive applications such as cognitive healthcare, smart transportation, smart city, and smart home can be implemented. Each and every application has their own architecture, hardware, software requirements and some challenges.

Cognitive computing defines technological platform which may combines reasoning, machine learning, Natural Language Processing, speech vision, that imitates human brain and tries to achieve the accuracy in human decision making. Some of the top companies who implementing the cognitive computing are Expert system, IBM Watson (Y.Chen et al., 2016), Cognitive scale, Numenta and etc.

Figure 1.

Figure 2. Cognitive Computing architecture

Expert system software which associates language and technology to transform it into use unstructured content. IBM Watson influences deep content analysis and evidences to improve decisions, to optimize the results. It practices a set of technologies such as hypothesis generation, evidence-based learning and influence natural language. Numenta is a technology based on machine learning and neocortex theory. The application of Numenta can be widely used in anomaly detection on server and applications, classification of NLP, models stock price, stock volume and etc.

As shown in Figure 1 cognitive computing frameworks are Tensor flow, Pytorch, Deeplearning4J, the Microsoft cognitive toolkit/CNTK, Keras, ONNX, MXNET, CAFFE. The functionality of each framework is explained as follows:

i)   Tensor flow: Tensor flow is a standard deep learning framework established by Google Brain team. Tensor flow provisions languages such as Python, C++, and R to generate deep learning models with additional libraries. Some of the use case of Tensor flow is Good text translation with NLP, classification of text, speech recognition, image recognition, fingerprint recognition, handwriting recognition, weather forecasting and etc. Its highly featured as robust and GPU support.

ii) Pytorch: Pytorch is a scientific based company which can support all machine learning algorithms. In addition is a supportive framework of deep learning which can be used in applications such as Twitter, facebook and etc. Pytorch can be run on Python and we can build our own deep learning models using this Pytorch in python.

iii) Deeplearning4J: Deeplearning4J is a deep learning library for Java Virtual Machine and it can be supported for Scala, Clojure, and Kotlin. When compare about the efficiency Deep learning4J is implemented in Java so it is more efficient than implanting in Python language. It can be used in applications such as mage recognition, fraud detection, text-classification and etc.

iv) Microsoft cognitive toolkit/CNTK: Microsoft cognitive toolkit/CNTK which can deliver higher performance and scalability as associated to toolkits such as Theano or TensorFlow. It can provision both RNN and CNN and it can be used in applications such as image recognition, handwriting recognition, and speech recognition.

v)   Keras: Keras can be written in python and it can support both CNN and RNN that are accomplished and run on both TensorFlow and Theano. Classification, text generation, and tagging are the primary use of cognitive computing.

vi) ONNX: Open Neural Network Exchange is established by Microsoft and Facebook as an open-source deep learning ecosystem. It is a deep learning framework that allows developers to change easily among platforms. It may act as providers for Tensor flow, CoreML, Keras, and Sci-kit Learn. It is widely

5

used in application such as handwriting recognition and optimization. It's having the greater advantage of flexibility and interoperability.

vii) MXNET: MXNET is a best deep learning-based framework which may runs on languages such as Python, R, C++, and Julia. To support for non- linear scaling efficiency this MXNET exploits the hardware as the greater maximum extent. It is having the greater of advantage of providing distributed training.

viii) CAFFE: CAFFE is a deep learning-based framework which could supported on languages or interfaces such as C, C++, Python, and MATLAB. The greatest advantage of Caffe's is speed. It can be processed more than sixty million images on daily with a single Nvidia K40 GPU. It can be mainly applicable in image recognition and vision recognition.

The following are the use of cognitive computing architecture in various applications.

COGNITIVE COMPUTING & INTERNET OF THINGS

Internet of Things relatively gather information and data from observed, supervised objects via through perception (input collection through sensors) equipment such as Radio Frequency Identification, wireless/ wired sensor, satellite location, observations via Wi-Fi, fingerprint and etc. At first the Internet of Things distributes the information via network. At last the method information will be progressed using machine learning, deep learning, cloud computing, data processing, pattern recognition to produce the most accurate and high precision decision making. Internets of Things finely understand about the input perceived data from the environment and also the information passed through it. Because of the prominence and most huge utilization of Internet of Things in day to day situation show the ways of delivering enormous measure of information which will give critical data hotspot for comprehension of cognitive computing. In the perspective on new part of computing technique, this cognitive processing will furnish a method for involving with higher energy productivity for information discernment and Internet of Things.

COGNITIVE COMPUTING AND BIG DATA ANALYSIS

Enormous information examination is the piece of the monstrous information esteem series which may focuses on changing over crude got information into a reliable usable asset that are promptly reasonable for investigation. Numerous scientists have talked with on key partners of different little and huge organizations and academicians.

They have revealed that there is serious areas of strength for a between cognitive processing and enormous information examination. It implied by associating people bid information thinking. Concerning applications on the off chance that the amount of data turns out to be high, it falls on large information issue which is various leveled as profound learning. The extremely at generally level of cognitive and enormous information examination is the uneasiness about advancement in substance life and climate. The second all things considered stage is following profound practices and the last however third level is restless with the significance of human life. The cognitive computing has an inclination of double-dealing of a rear entryway that is lightest and more reasonable than large information examination. This may find comprehensiveness and significance in scope of information and following acquiring cognitive knowledge, which develop and forms the huge information examination to not exclusively to utilize "brute computing force".

COGNITIVE COMPUTING AND CLOUD COMPUTING

Cloud computing and Internet of Things can be given cognitive computing programming premise and equipment premise. In most specifically huge information examination gives different strategy and suggestion for deciding and tracking down imaginative possibility and new worth in information.

COGNITIVE COMPUTING AND REINFORCEMENT LEARNING

These ordinary learning techniques are not skilled to get together the necessities of manageable advancement in knowledge of machine. Thusly, support learning is be changed over into a hot investigate branch in the field of machine learning. Reinforcement gaining can gain from the climate and ponder conduct.

COGNITIVE COMPUTING AND DEEP LEARNING

Rational Method and Perceptual Method

Truth be told, the kid doesn't have a clue about the idea of side or point yet the individual in question can in any case perceive a square. The strategy where the youngster perceives a square is perceptual technique or instinct. At the point when a kid figures out how to perceive a square, perceptual technique is utilized to lay out the planning connection among figure and idea, learning highlights with profound learning model can be seen as reproduction of this strategy. At the point when picture characterization is directed with profound learning, the perceptual strategy for human for picture acknowledgment is reproduced. The planning connection between the pictures and the grouping results is gotten through learning a lot of picture information. Coming up next are the moves toward do digital assault expectation.

1. Predicting the digital assault utilizing the cognitive man-made consciousness.
2. The most likely assault with the most noteworthy likelihood is

$P(J)=0.467$ WITH Doc 46.7%

3. the most likely time when an all the while digital assault at its most noteworthy, to be specific at time to duplication.
4. The Most likely assault classification with the most elevated assault likelihood which would happen whenever.
5. Hardening the PC organization or framework to expect the assaults from Exploit, fuzzer and conventional.
6. CAI based framework with information it gets cooperating with the peculiarity for 10 perceptions can convey three expectations, for example, most likely append class which would do assaults whenever later on.
7. The Prediction or assessment for the most plausible assault class with the most noteworthy likelihood of assault in what was in store was not difficult to do on the grounds that the quantity of occasions of such assault class is the most noteworthy among others.
8. The forecast or assessment will be a test on the off chance that the quantities of occasions of all assault classifications are practically like each other.
9. The Distinguishing component of CAI to other AI techniques in CAI acquires information by performing forward learning, to be specific by advancing by association

SMARTER COGNITIVE AND CYBER SECURITY WITH ARTIFICAL INTELLIGENCE:

Cyber security is a fast-evolving field because of advanced technologies which have increased the number of cyber criminals and cyber-attacks in the last decade. This study concentrates on cyber threats and how artificial intelligence is applicable in solving cyber security threats. This study I a review of existing data and analysis of empirical findings

Figure 3. SVM & AI Network

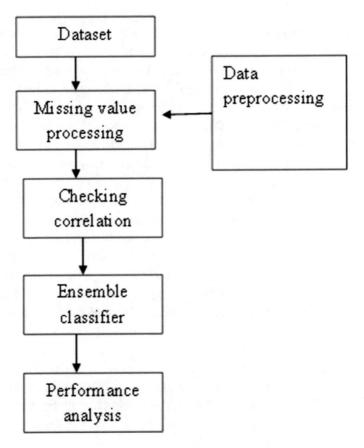

This block illustrates a example of support vector machines and Artificial intelligence network used for cyber security is a health care system, Artificial intelligences a based smart forecast of clinical illness random approaches and a investing strategy requires intelligent data mining.

A foundation area of research in contemporary artificial intelligence is the creation of independent agents is facilitating interaction efficiently with other precipitator to help in medical diagnosis. Artificial Intelligence techniques such as ANN, ML and DLN can monitor and record explicit learning events.

This study has shown that the traditional intrusion detection technologies and applications are less likely to mass trends of threats. These technologies use more elegant neuron approaches just like human brain to separate that threats from unlikely source and implement effective preventive measures.

COGNITIVE INTERNET OF VECHILES

Traffic the executives' framework (Khondokar et al., 2019) was imagined since the year 1990 by joining different sensors and different method of uses under the technical evolution of clever vehicle framework (ITS). At prior stage mechanical undertaking advanced with helping correspondence development named as Co-operative intelligent transport system (CITS) which cognitively integrates information and communication technologies (ICT) with transportation foundation. With the advancement of the Internet of Things (IoT) in mid 2010s, vehicles are being associated with web prompts one more mechanical advancement named as Internet of Vehicle (IoV)

The Necessity for applying Machine learning (ML), Neural Network (NN), Deep Learning, Artificial Intelligence that can assume command over a Wheel which can empower free driving, coming about to thought of Cognitive Internet of Vehicles (CLOV). This study targets giving an outline of the development of Cognitive Internet of vehicles and its mechanical mechanization in their activity yet not completely by any means. For the most part every one of the computerized vehicles are furnished with the number with number of sensors, cameras, Lidar, Radar and so on that gather's crude information from outer climate. This information that fills in as the contribution to the refined framework programming which is utilized in vehicle to choose the particular courses of activities, for example, path changing, speed increase and surpassing of vehicles. These vehicles utilize remote organizations to make advances to make connections with the gadgets worked in vehicles itself that is ready sensors and outside the vehicle. Vehicles to vehicle correspondence or vehicle to infrastructure (v2i). The Internet of vehicles consequently setting out a freedom to apply current mechanical advancements to apply information, for example, AI and computerized reasoning

Detecting and Participation is the layer 1 of the construction addresses every one of the mechanically advanced elements that are fit for detecting and conveying and furthermore answerable for collaborating with the transportation systems Layer in clov is liable for network-based correspondences among various transportation substances focusing on the vehicle related information procurement. Crash cautioning frameworks, path evolving frameworks, journey control, self-leaving framework are model that work utilizing the idea of bury vehicle communication. Edge figuring and Data Pre-handling layer structures on top of Vehicular cloud to help all kind of computing administrations at the edge of organization. This layer is additionally liable for offering continuous types of assistance to the taking part specialists

Cognizance and Control layer has number of safety and confidential issues distinguished which are for the most part connected with security issues and dangers. A portion of the security gives that exist in cloud worldview are Multitenancy,

Access, Availability, Misuse of cloud computing, Transborder information stream and information expansion and Trust. Application layer is the layer where security (M.Chen et., 2017) and confidential issues are subject to all low layers . The primary security worry in this layer can be postpone in arriving at information to applications.

This examination researches and centers around the security and confidential issues recognized alongside proposed engineering layer by layer. This will assist on the move with blurring based transportation, giving all security administrations expected via independent vehicles.

COGNITIVE COMPUTING

A Set of advancements that considers more prominent adaptability with greater and less unbendingly organized informational collections and which permits PCs to work with things like text and pictures in a way recently related exclusively with humans. Artificial and cognitive advances are supposed to have far and wide use at virtually all organizations in all areas of the economy. Media organizations are extraordinarily tested by the weakening of customary plans of action by the multiplication of new dispersion channels and by the need to conclude what kinds of new satisfied they ought to extend assets delivering and how they ought to best coordinate the substance with the rising divided client base

Media transmission is a south Florida based telecom suppliers whose foot print spans the globe. The artificial robotized arrangement decreases the activity to the division and with a lot more prominent unwavering quality. Mechanical and cognitive robotization imparts a ton to regular cousin in the production line.

Original robotizations are as of now generally utilized in business, to a great extent as a result of their overall effortlessness one of the lobby signs of man-made reasoning is permitting PCs to manage unstructured information as effectively as they do with data flawlessly organized in lines and sections. The capacity to get to cognitive experiences should be circulated justly all through the pieces of the association where are they generally required

Cognitive commitment has customary client care works regularly directed to a call habitat are rather given through computer. Artificial intelligence mechanical advances particularly in fields like normal language. Regular language handling includes the unprecedented complex undertakings of getting a PC to do what individuals achieve without exertion sorting out what somebody implies when they mean something. Fruitful utilization of cognitive advancements includes significantly more than dashing on a few seller items with little respect for associations funda cognitive construction and cycles

OTHER APPLICATIONS

Cognitive Computing and Robot Technology

New age of robot framework will mimic people from additional perspectives in future, particularly, there ought to be a sort of organization connection where robots and people coincide as one and make their separate benefits reciprocal to one another.

EMOTIONAL COMMUNICATION SYSTEM

Consequently, feeling perception is a significant use of cognitive computing. Furthermore, this is another human-machine connection technology. Emotional data cannot exclusively be perceived yet can likewise be sent in significant distance. In the meantime, a profound correspondence convention is proposed by the framework thinking about the ongoing necessities and guaranteeing the dependability of close to home correspondence.

CLINICAL COGNITIVE SYSTEM

A clinical cognitive framework can be used to aid finding and to settle on choice involving different sorts of information and items to take proper operation. The multidisciplinary combination of innovations, for example, AI, AI and normal language handling can empower the cognitive processing to decide the mode and the connection of a sickness from information.

Limitations of Cognitive Computing

Limited Analysis of Risk

The cognitive frameworks come up short at breaking down the gamble which is absent in the unstructured information. This incorporates financial variables, culture, worlds of politics, and individuals. For instance, a prescient model finds an area for oil investigation. Yet, in the event that the nation is going through an adjustment of government, the cognitive model ought to think about this component. Along these lines human mediation is funda cognitive for complete gamble examination and ultimate choice making.

Meticulous Training Process

Initially, the cognitive frameworks need preparing information to totally grasp the cycle and get to the next level. The arduous course of preparing cognitive frameworks is in all likelihood the justification for its sluggish reception. WellPoint's monetary administration is confronting what is going on with IBM Watson. The most common way of preparing Watson for use by the backup plan incorporates assessing the text on each clinical approach with IBM engineers. The nursing staff continues to take care of cases until the framework totally figures out a specific ailment. In addition, the perplexing and costly course of utilizing cognitive frameworks exacerbates it.

OngoingResearch in Cognitive Computing

1. Cognitive computing Research: From deep learning to higher machine learning

Most of the industries emphasis on enlightening the efficiency and influence of Deep Learning technology and products inflowing the market currently, it is charming vibrant that Deep Learning approaches obligate their limits and thus the Artificial Intelligence techniques need novel inventions to obtain more human-like cognitive capabilities. Novel tools are desired to address real-world challenges to result novel areas with slight supplementary data, system adaptation done by continuous learning. To address these challenges, Cognitive Computing Research is going on at Intel Labs to drive Intel's innovation with the combination of machine intelligence and cognition.

2. Cognitive Computing Market 2021 – Trends, Growth & Forecast Research Report Till 2026

The as of late article shows that "Worldwide Cognitive Computing Market Growth 2021" by Decision Databases, covering fragments examination, provincial and nations level examination close by with primary organizations working on the lookout. Also, the report has fascinated on market size, offer, drifts, and anticipated to 2026. Extra, the report incorporates impact examination along the side with industry degree and pay.

CONCLUSION

Cognitive computing, it empowers machines to essentially think and acquire the identical way that humans do. It encompasses self-learning algorithms that depend on countless progressions and data, which could support and simulate the approach that a human intelligence works. In parallel it has a goal to support humans in their routine responsibilities and decision making in short of essentially exchanging human effort. This chapter provided the view on architecture of cognitive computing in various technologies such as 1. Internet of Things 2. Deep Learning 3. Cloud Computing. Later we explained the few applications of Cognitive- Artificial Intelligence based applications. At last, we have given some emerging topics where some fruitful ongoing research in Cognitive Computing.

REFERENCES

Chen, Yang, Hao, Mao, & Kai. (2017). A 5G cognitive system for healthcare. *Big Data Cognit. Comput.*

Chen, M., Hao, Y., Hu, L., Huang, K., & Lau, V. (2017, December). Green and mobility aware caching in 5G networks. *IEEE Transactions on Wireless Communications*, 6(2), 8347–8836. doi:10.1109/TWC.2017.2760830

Chen, M., Herrera, F., & Hwang, A. K. (2018, April). Cognitive computing: Architecture, Technologies and Intelligent Applications. *IEEE Access: Practical Innovations, Open Solutions, 6*, 19774–19783. doi:10.1109/ACCESS.2018.2791469

Chen, M., Miao, Y., Hao, Y., & Huang, K. (2017). *Narrow band Internet of Things* (Vol. 5). IEEE Access.

Chen, Y., Argentinis, J. E., & Weber, G. (2016). *IBM Watson: How cognitive computing can be applied to big data challenges in life sciences research* (Vol. 38). Clin Therapeutics.

Hasan, Kaur, Hasan, & Feng. (2019). Cognitive Internet of Vehicles: Motivation, Layered Architecture and Security Issues. *International Conference on Sustainable Technologies for Industry 4.0.*

Chapter 2
Use of Socio–Cognitive and Affective Computing to Teach Emotions to Autistic Children

R. Muthuselvi
Kamaraj College of Engineering and Technology, India

G. Nirmala
Kamaraj College of Engineering and Technology, India

ABSTRACT

Learning aids in the development of attitude. It encourages the individual to learn new skills. It is critical to master three learning domains: cognitive, affective, and psychomotor. The cognitive computing system instantaneously processes data and gives solutions to questions. Affective computing is the development of tools that can recognise, understand, examine, and replicate human brains. Communication and behaviour are impacted by autism spectrum disorder (ASD), a developmental disease. ASD leads to have difficulty in interacting with society and communicating with society. It states that people with ASD have 1) difficulty in conversation and contact with other people, 2) symptoms that interfere with the person's ability to function normally in the society, and 3) restricted interests and repetitive behaviours. In the chapter, a computer-based model is developed for various emotions, facial expressions, and voice and body language. The aim is to develop a computer-based model that supports the autistic children to understand emotions and express their feelings.

DOI: 10.4018/978-1-6684-3843-5.ch002

Copyright © 2023, IGI Global. Copying or distributing in print or electronic forms without written permission of IGI Global is prohibited.

In the proposed work, computer based model can be developed for various emotions, facial expressions, voice and body language. The computer based model can be repeatedly used to teach the autistic children which will assist the autistic children in understanding the emotions. The self-learning algorithms, algorithms for pattern recognition can be used to know the emotions, understand the facial expressions for various emotions. Natural language processing can be used to know the words related to the emotions.

The aim is to develop a computer based model which supports the autistic children to understand the emotions and express their feelings.

1. INTRODUCTION

Affective computing (Jose Maria Garcia-Garcia et al., 2021) is knowledge from different disciplines of cognitive science, and psychology. It is the advancement of technologies that can perceive, analyze, process, and simulate human emotions. It is nothing but understanding of how emotions affect behavior and thought processes in people. One of the key components of affective computing is the detection of user emotions. Technologies for affective computing are designed to understand and react based on human responses.

Emotion Artificial Intelligence is another name for affective computing. It gives computers the ability to recognize, analyze, and imitate human moods and emotions. It draws inspiration from a variety of disciplines, including psychology and cognitive science. This knowledge allows a system to process the data gathered from countless sensors to evaluate the emotional state of an individual.

Social computing connects human social behaviour to computational systems. It is built on using software and technology to create or recreate social situations and social remedies. To determine how people perceive things and use accessible social knowledge, a variety of methodologies and methods are used. These techniques and strategies are together known as social cognition. Emotion processing, social processing, mentalising, and attribution style/bias are the four important key domains in social cognition (Pinkham et al., 2014).

2. MAIN DOMAINS OF SOCIAL COGNITION

2.1 Emotion Processing

Observing and using emotions is very crucial in Social cognition. Higher degrees of perception involve supervision and flexible emotion, whereas lower levels include identifying emotion in speech or faces.

2.2 Social Perception

Social perception is nothing but interpreting and construing social signs given by others. It contains information on social regulations and how these guidelines impact how people behave.

2.3 Mentalising or Theory of Mind

Mentalising is the capacity to assess the intents, attitudes, and/or beliefs of others by drawing conclusions about their mental states.

2.4 Attributional Style

The method by which people justify the origins of or attempt to make sense of, social occurrences or interactions is attributional style. How we interpret, store, and use knowledge about other people and social contexts is referred to as social cognition.

Emotions are mental states triggered by neurophysiological changes, differently linked to feelings, thoughts, behavioral reactions, and a level of desire or dissatisfaction that are informative and motivating primarily due to their experience or emotional component. Emotions make up the main source of motivation for mental processes and outward behavior.

It emphasizes that emotions are not just something you can experience.

2.5 Emotions Theory

The evolutionary theory of emotion, advanced by Charles Darwin, contends that feelings are situation-adaptive and increase our chances of enduring hardship. Human attention, learning, perception, memory, problem-solving, and reasoning are just a few of the cognitive functions that emotion significantly affects. Emotion has a significant influence on attention, particularly in altering attentional selectivity and in driving behavior. The complete picture of emotions encompasses cognition, physical experience, action, and limbic/pre-conscious experience. Let's examine several aspects of emotion in more detail.

2.6 Subjective Experience

While experts agree that some fundamental universal feelings are shared by everyone everywhere, regardless of education or culture, academics also hold that experiencing emotions can be very subjective.

2.7 Physiological Response

The body's automatic reactions to a stimulus are known as physiological responses.

2.8 Behavioral Response

Basic complicated physiological changes that result from stress are indicative of behavioral responses to stress.

3. OVERVIEW OF AUTISM

Abnormalities in brain structure or function cause Autism spectrum disorder (ASD) (Jose Maria Garcia-Garcia et al., 2021), (Rana El Kaliouby, 2006). ASD is a developmental disability which includes a broad spectrum of disorders. ASD affects the social skills, speech, movement, learning, cognition, mood and behavior of the people. In a recent study, it is shown that 1.5 to 2 percent occurrence of ASD is in the general under-18 population. It is also shown that one in hundred children suffers from ASD on an average worldwide. However, the exact count may differ across the various studies.

Autistic children find difficulties in communicating with others. It is said by healthline.com that around 25% to 30% of autistic children are minimally verbal or not able to talk. Some of the autistic children know how to perform social skills but they may have difficulties to perform the social skills. They often find it hard to converse with others and may misread social cues. They are not able to make friends and have difficulties in social communication. Autistic children have abnormal functioning in the main development areas such as social interaction, verbal and non-verbal communication, presence of repetitive and restricted patterns of behaviors.

Several studies suggest that persons with ASD are more prone to an increased risk of tautness disorders. If the children with ASD have tautness disorder, they may have more complex behaviors like aggressive and oppositional behavior, impairment in social functioning, and increased negative thoughts. It is very crucial to treat these behavioral problems in time; otherwise they can not come out of the disorder permanently. Any treatment will not provide a solution for this problem. In the case of children with tautness disorder and ASD, the assessment is difficult. Some symptoms tautness and ASD overlap which leads to the difficulty in assessment.

Tautness is said to have somatic and cognitive modules. Cognitive or mental constituents are accompanying with negative outlooks which leads to poor performance. Somatic modules are associated with the stimulation of autonomic nervous system (ANS).

This biological stimulation is usually experienced by increase in rapidity of breath, heart rate, and sweating of tributes. Electrodermal activity (EDA) (Rosalind W. Picard, 2009) is a moral extent of arousal from the sympathetic nervous system (SNS). EDA may be useful to understand the emotional response of a person who talks very less. Recently, more studies are done on tautness and its interaction with emotion comprehension and expression in ASD. Several physiological signs, such as heart rate, EDA, muscle activity, hyperactive, inattentive behavior etc provide information to understand the affective state of an individual. There are many ways of stating emotions through speech (words and tone), facial expressions, and body gestures. There are a variety of approaches which are useful for studying biological arousal and subsequent emotion processes.

Emotional processing is defined as the modification of memory structures that underlie emotions. Peter Lang developed a model for bio-informational processing. Jack Rachman worked on the concept of emotional processing. Both of these works are the basics for tautness reduction.

Emotions (Ekman, 1999), (Golan, 2010) are the interior felt reaction to a specific stimulus. Basically there are five reactions fear, joy, sadness, anger, and disgust. There are thousands of feelings. The cause is the circumstance, in which the emotions are instinctive response without the assistance of mental processing. Feelings are a mindful subjective experience of emotion. There are various ways to process the emotions. All forms of emotional processing require recognizing and feeling without judging oneself for having them.

Recently, EDA is included as an arousal measure in persons with ASD. Emotion eliciting pictures are the stimuli used to measure EDA data in many these studies and the use of those leads to discrepancy. Another stimulus is stress tasks, a set of authenticated procedures. However, EDA data is not adequate to understand the emotions experienced by offspring with ASD in their day-to-day life. The laboratory studies of EDA in ASD in response to stimuli fails to provide the required conceptualization of the day-to-day functioning of the individual. Hence, both of these types of investigations are indispensable. However, the garage measurements of EDA are not much useful for the study of emotions of children with ASD. Facial expressions have long been studied as a means of empathetic emotions. As shown by Fig 1, the analysis of facial expressions includes the following ordered steps: (1) recognition of the face, (2) mining of facial features, and (3) ordering of facial expressions.

Figure 1. Analysis of facial expressions

For human communication, facial expressions are very important. The face is able to express thoughts, ideas and emotions. These facial expressions of emotion such as anger, disgust, fear, happy, sad, surprise are expressed the same way by all people in the world. The Facial Action Coding System (FACS) (Cohn & Kanade, 2007) is an important technique that is used to study facial expressions. Whatever the classification method used, there are limitations in FACS. EDA trainings that are relevant to the emotion perception and production of facial expressions by persons with ASD are not able to provide a compromise. Trainings suggest that persons with ASD do not perform in labeling basic emotions. In emotion based trainings, one approach examines the emotion perception of persons with ASD. A second approach examines the facial emotion production of persons with ASD.

It is required to establish computational thinking of ASD to represent a set of skills related to Computer Science with EDA. Recently, more individuals are analyzed with ASD and it shows the importance of computational thinking.

Computational tools become essential which help in analyzing accurately. As depicted in Fig 2, the analysis includes logical organization of data, breaking problems into smaller pieces, pattern and model interpretation and designing and implementing algorithms.

4. TEACHING EMOTIONS TO AUTISTIC CHILDREN

The emotional response of the children with ASD can be recorded. A model can be developed for the recorded responses using affective computing approaches. Embodied conversational agents or robots can be engaged to express emotions (Graesser & D'Mello, 2011; Picard, 1997). Sameness and routines are the important characteristics of autistic children. Hence, unvarying, predictable interactions can

Figure 2. Computational thinking

be given by embodied conversational agents. Embodied Conversational Agents (ECA) are software based automata that can be used to assist children with ASD in emotional or other tasks. Agents are represented with a human audiovisual form whose appearance ranges from cartoon-like to photographic. ASD Children are able to communicate as much with an ECA as with a human psychologist using the same script (Black, Flores, Mower, Narayanan, & Williams., 2010). ECA supports the children with ASD to concentrate on the cognitive aspects of learning. Using robots is also as valuable as ECA for these children.

Recently, more focus is on the interplay between emotions and learning and the role of emotions in learning ((Cicchetti & Sroufe, 1976; Graesser & D'Mello, 2011; Kort, Reilly, & Picard, 2001). Research study on emotions and computing shows that a bewildering array of emotions is relevant to learning. It is also shown that negative emotions like frustration, confusion, boredom, etc., facilitate deep learning (Graesser & D'Mello, 2011). As a consequence, affective-based agents are incorporated in many tutoring systems (Mao & Li, 2010) and many traditional academic applications (Arroyo, Woolf, Royer, & Tai, 2009) by ITSs.

Affective Auto-Tutor is fully automated, affect-aware dialogue-based intelligent tutoring system (ITS). ITS interacts with the students in natural language. It is able to use adaptive dialog moves like a human tutor (D'Mello & Graesser, 2013). It identifies the emotions of students and uses the information to regulate their emotions

(D'Mello & Graesser, 2012). The performance of ITS is so good that good learning outcomes are shown by the students. It can be a better software to be used for children with ASD. The architecture of ITS is shown in Fig 3.

Figure 3. Intelligent Tutoring System

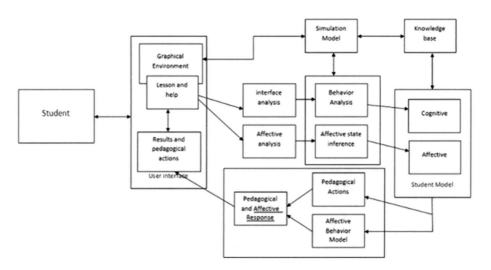

Agent-based intervention systems provoke empathy and the learner can experience and express different target emotional states. FearNot! (Fun with Empathic Agents to Achieve Novel Outcomes in Teaching) an agent-based system which is used to elicit emotion and teach children with ASD (Paiva et al., 2004) to regulate emotions. A bully, a victim, and a narrator are the three affect-based agents used by FearNot! to teach different coping strategies.

Agents had a model that represents the agent's own emotions and others' emotions. Agents had a parameter-based personality which included role-based (e.g., victim or bully) thresholds for experiencing different emotions. A function that recalculated the intensity of equivalent emotions was used. An action selection module was framed and it included action tendencies which were not planned and based on the agent's role and personality.

For children with ASD, embodied conversational agents can be used to improve learning (Bosseler & Massaro, 2003). With the help of ECAs, the children with ASD learn to recognize emotions in others and in themselves. When the children with ASD recognize emotions and do emotion storytelling tasks, the multi model data is collected by Rachel, a pedagogical emotional coach. He analyzes the data and finds a good result.

The children with ASD interact with the agent. A therapist can monitor the system in real time (Mower et al., 2011). Using Support vector machine classification; Rachel finds acceptable interactions from children with ASD in the context of emotional learning. Systems might distinguish different facial expressions and physiological signals in children with ASD. Systems might prompt them to report on their emotional experiences by matching their emotional experience to sample emotional faces. The self-recognition and expression of emotions can be thereby improved.

Recently, several studies have been done on the response of children with ASD to both humanoid robots and non humanoid toy-like robots. It is suggested by Diehl, Schmitt, Villano, & Crowell (2012) to utilize robotic systems to develop novel interventions and enhance existing treatments for children with ASD. It was shown by Kim and colleagues (2012) that children with ASD were able to speak more to an adult confederate when asked by a robot than when asked by another adult or by a computer. Duquette and colleagues (2008) found that children paired with a robot had greater increases in shared attention than did those paired with a human. Improved enhancements in interactions by the children with ASD were reported by Goodrich and colleagues (2011). They also reported that with robot acting contingently during an interaction with a child with ASD, the social interaction was very much improved.

5. CONCLUSION

In this work, a systematic review has been provided on affective computing and socio cognition in general. The various domains of socio cognition are discussed. The difficulties faced by the autistic children are explained. In specific, the problem of communication skill and understanding emotions faced by the autistic children are dealt in the work. The various approaches that use the recent computing technologies for finding the solution for these problems are discussed. The impacts of these technologies are also dealt in detail.

REFERENCES

Affective Computing and Autism. (2006). *Massachusetts Institute of Technology*.

Baraka, K., Marta Couto Francisco, S. M., & Paiva, A. (2022). *Sequencing Matters*. Investigating Suitable Action Sequences in Robot-Assisted Autism Therapy Frontiers in Robotics and AI, doi:10.3389/frobt.2022.784249

Beverly, W. W. B. I. A., Dragon, T., & David, C. R. P. (2009). Affect-aware tutors: Recognising and responding to student affect. *International Journal of Learning Technology*, *4*(3/4), 129–164. doi:10.1504/IJLT.2009.028804

Bosseler, A., & Massaro, D. W. (2004). Development and Evaluation of a Computer-Animated Tutor for Vocabulary and Language Learning in Children with Autism. *Journal of Autism and Developmental Disorders*, *33*(6), 653–672. doi:10.1023/B:JADD.0000006002.82367.4f PMID:14714934

Chaidi & Drigas. (2020). Autism, Expression, and Understanding of Emotions: Literature Review Article. *International Journal of Online Engineering*, 125–147.

Cicchetti, D., & Sroufe, L. A. (1976). The relationship between affective and cognitive development in Down's syndrome infants. *Child Development*, *47*(4), 920–929. doi:10.2307/1128427 PMID:137105

Cohn, J., & Kanade, T. (2007). Automated facial image analysis for measurement of emotion expression. In J. A. Coan & J. B. Allen (Eds.), The handbook of emotion elicitation and assessment (pp. 222–238). Oxford.

D'Mello, S., & Graesser, A. (2011). The Half-Life of Cognitive-Affective States during Complex Learning. *Cognition and Emotion*, *25*(7), 1299–1308. doi:10.1080/02699931.2011.613668 PMID:21942577

D'Mello & Graesser. (2012). AutoTutor and affective AutoTutor: Learning by talking with cognitively and emotionally intelligent computers that talk back. *The ACM Transactions on Interactive Intelligent Systems*.

Diehl, Schmitt, Villano, & Crowell. (2012). The Clinical Use of Robots for Individuals with Autism Spectrum Disorders. *Critical Review*. Advance online publication. doi:10.1016/j.rasd.2011.05.006

Ekman, P. (1999). *Basic emotions. In Handbook of cognition and emotion*. Wiley.

Pichard, R. W. (2009). Future affective technology for autism and emotion communication. *USA Phil. Trans. R. Soc. B*, *364*, 3575–3584. doi:10.1098/rstb.2009.0143

Garcia-Garcia, Penichet, Lozano, & Fernando. (2021). *Using emotion recognition technologies to teach children with autism spectrum disorder how to identify and express emotions*. Academic Press.

Golan, O., Ashwin, E., Granader, Y., McClintock, S., Day, K., Leggett, V., & Baron-Cohen, S. (2010). Enhancing emotion recognition in children with autism spectrum conditions: An intervention using animated vehicles with real emotional faces. *Journal of Autism and Developmental Disorders*, *40*(3), 269–279. doi:10.100710803-009-0862-9 PMID:19763807

Graesser, A., & D'Mello, S. K. (2011). Theoretical perspectives on affect and deep learning. In R. A. Calvo & S. K. D'Mello (Eds.), *New perspectives on affect and learning technologies* (Vol. 3, pp. 11–21). Springer. doi:10.1007/978-1-4419-9625-1_2

Holdnack & Saklofske. (2019). WISC-V and the Personalized Assessment Approach. WISC-V.

Jonsson, U., Choque Olsson, N., & Bölte, S. (2016). Can findings from randomized controlled trials of social skills training in autism spectrum disorder be generalized? The neglected dimension of external validity. *Autism*, *20*(3), 295–305. doi:10.1177/1362361315583817 PMID:25964654

Kort, B., Reilly, R., & Picard, R. W. (2001). *An affective model of interplay between emotions and learning: Reengineering educational pedagogy—Building a learning companion.* Paper presented at the International Conference on Advanced Learning Technologies, Madison, WI. 10.1109/ICALT.2001.943850

Mao, X., & Li, Z. (2010). Agent based affective tutoring systems: A pilot study. *Computers & Education*, *55*(1), 202–208. doi:10.1016/j.compedu.2010.01.005

Nisiforou, E. A., & Zaphiris, P. (2020). *Let me play: Unfolding the research landscape on ICT as a play-based tool for children with disabilities. Universal Access in the Information Society, 19(1).*

Philip, R. C. M., Whalley, H. C., Stanfield, A. C., Sprengelmeyer, R., Santos, I. M., Young, A. W., Atkinson, A. P., Calder, A. J., Johnstone, E. C., Lawrie, S. M., & Hall, J. (2010). Deficits in facial, body movement and vocal emotional processing in autism spectrum disorders. *Psychological Medicine*, *40*(11), 1919–1929. doi:10.1017/S0033291709992364 PMID:20102666

Pinkham & Badcock. (2020). Assessing Cognition and Social Cognition in Schizophrenia & Related Disorders. A Clinical Introduction to Psychosis, 201-225.

Provost, E. M., Black, M., & Flôres, E. L. (2011). Rachel: Design of an emotionally targeted interactive agent for children with autism. *IEEE International Conference on Multimedia and Expo (ICME).*

Ramos-Aguiar, L. R., & Álvarez-Rodríguez, F. J. (2021). Teaching Emotions in Children With Autism Spectrum Disorder Through a Computer Program With Tangible Interfaces. IEEE Revista Iberoamericana, 16(4), 365 – 371.

Rangasamy, S., D'Mello, S. R., & Narayanan, V. (2013). *Epigenetics, Autism Spectrum, and Neurodevelopmental Disorders. Journal of the American Society for Experimental NeuroTherapeutics*, *10*(4). Advance online publication. doi:10.100713311-013-0227-0 PMID:24104594

Chapter 3
An Emotion-Aware E-Learning System Based on Psychophysiology

Jerritta Selvaraj
Vels Institute of Science, Technology, and Advanced Studies, India

Arun Sahayadhas
Vels Institute of Science, Technology, and Advanced Studies, India

ABSTRACT

Computer-based learning and training has increased over the past few decades and has become the norm of today. This allows students to learn from the comfort of their home at their convenient time and pace. The learning materials can be accessed at any time and from anywhere making it easier for students. Recent studies indicate that negative emotions such as anger, frustration, confusion, and boredom inhibit the learning process and positive emotions such as excitement and enthusiasm support the learning process. Many students complain of boredom and similar negative emotions, which deteriorate the rate of learning. This research tries to integrate the emotion recognition system to help students during the learning process. Appropriate coping mechanisms can be integrated into the e-learning system to keep the students engaged and alert in the process. Emotions can be identified using wearable sensors and appropriate learning activity or breaks or physical activity can be given to students based on the emotion experienced by the students.

DOI: 10.4018/978-1-6684-3843-5.ch003

Copyright © 2023, IGI Global. Copying or distributing in print or electronic forms without written permission of IGI Global is prohibited.

1. INTRODUCTION

The blooming of the knowledge era coupled with the pandemic and reliance on the Internet has increased the number of students and employees to rely on computer based learning methods such as Massive Open Online Courses (MOOCs), Intelligent Tutoring Systems (ITS) etc., (Imani & Montazer, 2019). The number of courses offered via MOOCs have compounded over the years as in Figure 1. Various modalities such as synchronous and asynchronous e-learning are proposed by faculty and researchers to improve the quality of teaching and learning.

These have brought world-class education into our homes and have open end wide avenues to learn and excel in the desired scope and subject of study. Students, researchers, faculty members and employees of multi national companies equip through the various e-Learning platforms that are available (Hökkä et al., 2020). The endless possibilities have made MOOCs an influential parameter in the higher education spectrum(Muñoz et al., 2020). Governments are devising policies to inculcate e-Learnings into their education systems. The SWAYAM/NPTEL courses in India provides opportunities to learners at all levels to learn anywhere and anytime (Chauhan & Goel, 2017). Researchers have indicated that the learning outcome is improved with the skills and techniques learnt by students online. The idea of life-learning is also promoted through the various e-Learning and web based courses (Alhazzani, 2020).

Synchronous learning enables all learners to connect online at the same time as an online classroom approach replicating a traditional classroom on the screen. These scheduled classes are found to be more collaborative compared to the self-paced courses or asynchronous e-learning is that is largely dependent on the learner (Imani &Montazer, 2019; Min & Nasir, 2020). Learners are required to be self-regulated, guided by metacognition, plan, set goals, self- monitor, self-instruct and self-reinforce to be successful(Min & Nasir, 2020). Though they are widely used by many learners, they have also reported high levels of boredom, frustration and anger (Stephan et al., 2019). Skills such as attention, learning, memory, reasoning and problem solving are affected by the human cognitive process (Tyng et al., 2017). The low retention and success rates of MOOC compared to the enrollment rate is also a matter of concern (Crane & Comley, 2021; Hamann et al., 2021).

As, learning is a cognitive process that is highly influenced by emotions, many of the recent researchers are focusing on finding the correlation between emotions and learning. Positive emotions enable faster learning, better retention capabilities and tends to widen the cognition, attention and action and engagement of learners (Imani &Montazer, 2019; Muñoz et al., 2020). The attentional and motivational components of emotion have been linked to enhanced learning and memory. Designers of educational courses should work to maximize learner engagement to improve learning and long-term retention of the material (Shen et al., 2009). Understanding the need of emotion aware systems in learning, Muñoz et al., suggests to integrate emotion – aware systems and learning analytics to enhance the engagement and learning using e-Learning systems (Muñoz et al., 2020).

Emotion Recognition systems are widely used in various fields of Human Computer Interaction (HCI). Emotions are recognized using behavioural (facial actions, gestures, speech etc.,) and physiological methods (Electroencephalogram (EEG) signals from the brain, Electrocardiogram (ECG),signals from the heart, Electromyograph (EMG) signals from various muscles, Skin Conductance (SKC) etc.) (Jerritta et al., 2011). Physiological signals being an activity of the Autonomous Nervous System (ANS) reflects the experienced emotions in various ways. For e.g, the emotion of fear leads the ANS to provoke a stress response where activation of special types of hormones into the bloodstream causes an increase in skin sweat, heart rate, and breathing frequency and the mouth becomes dry (Imani & Montazer, 2019). The socially masked and unexpressed emotions can also be identified using psychophysiology based methods.

This book chapter presents a methodology to integrate the emotions detected from the Electrocardiogram (ECG) and Electromyograph (EMG) signals into an e-Learning module that would suggest the learner to perform an appropriate activity to normalize an emotion. The various steps in the detection of emotion ECG and EMG are discussed in detail. The GUI for e-Learning using the emotions recognized using ECG and EMG are also presented.

2. METHODOLOGY

The development of an emotion aware e-Learning system should contain an efficient emotion detection module that needs to integrated into an E-Learning system so that the learner can recognize their emotions and follow the suggestion specified by the system. This sessions first describes the developed emotion recognition system and later the integration of the developed emotion recognition system into the E-Learning environment

2.1 Emotion Recognition System using ECG and EMG

As both ECG and EMG signals can be obtained in an non-intrusive manner, it was used to detect the emotions of the subject. Figure 2 illustrates the methodology used for developing the Emotion Recognition algorithm. In this work, six emotional states namely Happiness, Sadness, Surprise, Disgust, Fear and Neutral were identified from ECG and EMG. The emotion 'anger' was omitted as it was difficult to induce it in a controlled environment.

2.1.1 Emotion Elicitation Protocol

Emotional data is one of the important requirements to develop a system that can recognize the emotions. This makes it imperative to induce emotions in the subjects. Developing a protocol that can induce the emotions in the subjects depends on various factors such as, the method of emotion elicitation (elicited by the subject or by an event), the experimental setting (laboratory or natural expression), the nature of elicitation (a strong emotion or a simple feeling), the method of recording of data (known or unknown to the subject) and the subjects knowledge of the purpose of the experiment (Picard et al., 2001)

Different methods such as recall paradigm, audio, video etc., are used to elicit the emotions of subjects in various experiments (Jerritta et al., 2011). In this research work audio visual cues were presented to the subjects for eliciting the emotional states. More details of the protocol and emotion elicitation can be read in our earlier publication (Selvaraj et al., 2013).

2.1.2 Data Acquisition

The set-up of the data collection experiment is as shown in Figure 3. The participants watched the video on the project screen while comfortably sitting on a chair. Electrodes were connected to capture physiological data. Facial actions were also recorded. Consent was obtained from all the participants (parents, in case of children), before the data collection experiment was performed. The participants were briefed and requested to not suppress the emotions but be relaxed throughout the experimental process.

Sixty healthy volunteers in varied age groups participated in the experiment. The age group was categorized into three and included fifteen children belonging to the age group of nine to sixteen, thirty students with ages varying between eighteen and twenty five years and fifteen adults whose ages varied between thirty nine and sixty eight 68 years participated in the data collection experiments. The number of male and female participants were equal in all the groups and were recruited from a nearby school, university and parents of the children.

The emotional ECG data was acquired using the PowerLab data acquisition system with chart software, developed by AD Instruments, Australia. Three electrodes system with was used to capture the emotional ECG data. EMG data was collected using Bipolar electrodes on the Zygomaticus Muscle using PHYWE hardware and

'Measure' software. The position of electrodes is as indicated in Figure 4. ECG signals were sampled at 1024 Hz and EMG signals at 512 Hz.

2.1.3 Pre-Processing

Both the ECG and EMG signals acquired are prone to different types of noises and artifacts due disturbances in power line, movement, muscular artefacts etc., Power line interference and high frequency noised in ECG signals was removed by using 4th order Butterworth filter with a cut off frequency of 45 Hz. Baseline wander was removed using the Discrete Wavelet Transform (DWT) based algorithm (Bunluechokchai& Leeudomwong, 2010).

Superposition spikes, random changes in the amplitude and short term spikes were removed from the EMG signals using 128 point moving average window smoothing filter (Haag et al., 2004). The continuous piecewise linear trend was removed from the EMG signals to retain the emotional variations.

2.1.4 Feature Extraction

Obtaining the emotional information from the ECG signals play a major part in developing a machine learning algorithm that can identify the emotional states. In this work, Empirical Mode Decomposition based analysis was used to identify the emotional states as in Figure 5.

Hilbert Transform and Discrete Fourier Transform was applied to the instantaneous amplitude of Intrinsic Mode Functions (IMF) derived from the ECG and EMG signals after applying the Empirical Mode Decomposition algorithm. Statistical, Higher Order Statistical (HOS) and non-linear features were derived from the various IMF's and analysed.

The six statistical features as proposed by Picard (Picard, et al., 2001) for emotion recognition applications are specified in equations 1 through 6. HOS features such as skewness and Kurtosis was derived as in equations 7 and 8. Nonlinear features namely Hurst and Lyapunov exponent was calculated from the IMF's.

Assuming the signal X_n, $n=0,1,2,\ldots,N$ to be of zero mean,

Mean of raw signal

$$\mu_x = \frac{1}{N}\sum_{n=1}^{N}X_n \qquad (1)$$

Standard Deviation of raw signal

$$\sigma_x = \frac{1}{N-1}\sum_{n=1}^{N}\left(X_n - \mu_x\right)^2 \tag{2}$$

Mean of absolute values of first differences of raw signals

$$\delta_x = \frac{1}{N-1}\sum_{n=1}^{N-1}\left|X_{n+1} - X_n\right| \tag{3}$$

Mean of absolute values of first differences of normalized signals

$$\delta_x^{\text{Æ}} = \frac{1}{N-1}\sum_{n=1}^{N-1}\left|X_{n+1}^{\text{Æ}} - X_n^{\text{Æ}}\right| \tag{4}$$

Mean of absolute values of second differences of raw signals

$$\gamma_x = \frac{1}{N-2}\sum_{n=1}^{N-2}\left|X_{n+2} - X_n\right| \tag{5}$$

Mean of absolute values of second differences of normalized signals

$$\gamma_x^{\text{Æ}} = \frac{1}{N-2}\sum_{n=1}^{N-2}\left|X_{n+2}^{\text{Æ}} - X_n^{\text{Æ}}\right| \tag{6}$$

Skewness,

$$\vartheta_3^{\text{Æ}} = \frac{\sum_{n=1}^{N}\left(X_n - \mu_x\right)^3}{(N-1)\sigma_x^3} \tag{7}$$

Kurtosis,

$$\vartheta_4^{\text{Æ}} = \frac{\sum_{n=1}^{N}\left(X_n - \mu_x\right)^4}{(N-1)\sigma_x^4} - 3 \tag{8}$$

where μx represents the mean and σx the standard deviation of the signal Xn defined in equation 7 and equation 8 respectively.

2.1.5 Classification of Emotional States

Naive bayes, regression tree classifier, KNN and Fuzzy KNN classifiers were used to classify the emotional states. Three performance measures namely percentage Accuracy, Sensitivity and Specificity were used to study the performance of classifiers in classifying the emotional states,

$$\% Accuracy_E = \frac{Number\ of\ correctly\ classified\ samples_E}{Total\ number\ of\ samples\ tested_E} \times 100 \qquad (9)$$

$$Sensitivity_E =$$
$$\frac{Number\ of\ correctly\ classified\ samples_E}{Number\ of\ correctly\ classified\ samples_E + Number\ of\ incorrectly\ classified\ samples_E} \qquad (10)$$

$$Specificity_E =$$
$$\frac{Number\ of\ correctly\ classified\ samples_{AE}}{Number\ of\ correctly\ classified\ samples_{AE} + Number\ of\ E\ samples\ identified\ as\ AE} \qquad (11)$$

where the subscript E refers to the emotional states such as happiness, sadness, fear, surprise, disgust and neutral and subscript AE represents the classification of all the emotional states except E. The '*Number of incorrectly classified samples$_E$*' refers to the samples that are incorrectly classified as emotion E.

2.2 Integration of Emotion Recognition Algorithm into an E-Learning Environment

Researchers have correlated positive emotions to productivity and enhanced learning (Imani & Montazer, 2019). Once the emotions are identified, appropriate plan of action needs to be indicated to the students to improve their productivity. Table 1 indicates the proposed plan of action on identifying the emotions. As emotions are dependent on the user, the plan of action system, can be personalized according to the preference of the user

Table 1. Identified Emotion and Plan of Action

SI. No	Identified Emotion	Proposed Plan of Action
1	Neutral	No action, the student continues to study
2	Happiness	No action, the student continues to study
3	Sadness	Relax by listening to Instrumental Music. Recheck Emotion later
4	Disgust	Play a game to clear thoughts. Recheck emotion later
5	Surprise	Calm down and start studying after five minutes. Recheck emotion if needed.
6	Fear	Relax by listening to Instrumental Music. Recheck Emotion later

3. RESULTS AND DISCUSSION

This section discusses the results of the developed Emotion Recognition and the Prototype of Emotion Aware e-Learning system. The Graphical User Interface and its components are elaborated in detail.

3.1 Emotion Recognition using ECG and EMG

Figure 6 and Figure 7 demonstrates the removal of noises from ECG and EMG signals respectively. Though the emotions were collected in an closed environment, it is prone to the low frequency movement artefacts, which needs to be removed for proper recognition of emotional states. Similarly, the EMG signals were smoothed and trends were removed before further processing.

The pre-processed signals were decomposed using EMD algorithm and the derived IMF's are displayed in Figure 9 for the ECG signals. Hilbert Huang Transform (HHT) and Fourier transform were applied to the IMFS's. The statistical, HOS and non-linear features were extracted from the Instantaneous Amplitude signals obtained from HHT. The Instantaneous Amplitude and Frequency of the ECG signals are as in Figure 9. In the case of FFT, the features were extracted from the frequency domain of the IMF's.

The performance of the various features derived from the ECG signals after EMD, EMD followed by HHT and EMD followed by FFT are as indicated in Tables 2 through 4 respectively.

In all cases, it can be observed that the accuracy of statistical features using the FKNN classifier performed better with 81.79%, 81.98% and 86.72% accuracy in classifying the six emotional states. The sensitivity and specificity of the FKNN classifier with higher accuracy is as tabulated in Table 5.

Table 2. Accuracy of ECG features after EMD

Feature	Classifier	Neutral (%)	Happiness (%)	Sadness (%)	Fear (%)	Surprise (%)	Disgust (%)	Average (%)
Statistical Features	Regression Tree	76.11	81.01	55.11	71.91	49.40	84.75	69.72
	Naïve Bayes	35.56	50.84	20.45	49.44	47.02	55.37	43.11
	KNN (K=6)	70.00	76.54	55.68	79.78	53.57	84.75	70.05
	FKNN (K=6)	**80.00**	**85.47**	**73.30**	**85.39**	**76.19**	**90.40**	**81.79**
HOS Features	Regression Tree	28.89	18.99	23.30	24.72	22.62	12.99	21.92
	Naïve Bayes	38.33	25.14	3.98	3.93	71.43	0.00	23.80
	KNN (K=9)	25.00	21.79	21.59	32.58	33.33	25.99	26.71
	FKNN (K=10)	20.00	20.11	19.89	26.97	30.36	24.86	23.70
Nonlinear Features	Regression Tree	49.28	46.38	30.15	43.07	45.45	43.80	43.02
	Naïve Bayes	57.25	47.10	13.97	65.69	55.30	54.74	49.01
	KNN (K=7)	47.83	57.25	30.88	50.36	54.55	39.42	46.71
	FKNN (K=8)	42.03	49.28	28.68	47.45	47.73	39.42	42.43

Table 3. Accuracy of ECG features after EMD and HHT

Feature	Classifier	Neutral (%)	Happiness (%)	Sadness (%)	Fear (%)	Surprise (%)	Disgust (%)	Average (%)
Statistical Features	Regression Tree	84.06	73.91	55.15	73.72	77.21	83.21	74.54
	Naïve Bayes	42.75	26.81	34.56	36.50	75.00	71.53	47.86
	KNN (K=6)	77.54	78.26	63.24	78.83	52.94	82.48	72.21
	FKNN (K=6)	**81.88**	**81.88**	**74.26**	**85.40**	**75.00**	**93.43**	**81.98**
HOS Features	Regression Tree	26.09	26.09	20.59	21.17	30.88	27.01	25.30
	Naïve Bayes	1.45	38.41	8.09	24.82	72.06	26.28	28.52
	KNN (K=10)	25.36	23.19	25.74	18.25	37.50	22.63	25.44
	FKNN (K=7)	19.57	18.12	30.88	18.98	34.56	18.98	23.51
Nonlinear Features	Regression Tree	36.26	48.91	40.00	40.66	54.44	36.67	42.82
	Naïve Bayes	61.54	33.70	23.33	53.85	67.78	53.33	48.92
	KNN (K=7)	41.76	45.65	41.11	43.96	65.56	42.22	46.71
	FKNN (K=6)	41.76	47.83	30.00	43.96	52.22	34.44	41.70

Table 4. Accuracy of ECG features after EMD and FFT

Feature	Classifier	Neutral (%)	Happiness (%)	Sadness (%)	Fear (%)	Surprise (%)	Disgust (%)	Average (%)
Statistical Features	Regression Tree	88.41	81.88	83.09	84.67	74.26	93.43	84.29
	Naïve Bayes	14.49	42.03	44.85	38.69	33.82	62.77	39.44
	KNN (K=10)	82.61	84.78	63.24	79.56	51.47	82.48	74.02
	FKNN (K=6)	**89.86**	**89.13**	**86.03**	**84.67**	**77.94**	**92.70**	**86.72**
HOS Features	Regression Tree	24.64	34.78	26.47	20.44	61.03	40.15	34.58
	Naïve Bayes	8.70	65.94	19.12	35.04	75.00	24.82	38.10
	KNN (K=10)	23.91	46.38	22.79	27.01	70.59	32.12	37.13
	FKNN (K=9)	18.84	36.23	27.21	25.55	58.09	36.50	33.74
Nonlinear Features	Regression Tree	24.18	21.74	18.89	20.88	23.33	18.89	21.32
	Naïve Bayes	3.30	41.30	0.00	27.47	61.11	51.11	30.72
	KNN (K=9)	21.98	27.17	17.78	16.48	42.22	26.67	25.38
	FKNN (K=9)	23.08	23.91	14.44	27.47	30.00	27.78	24.45

Table 5. Performance Metrics of ECG based Emotion Recognition

	EMD	HHT	FFT
Accuracy	81.79%	81.98%	**86.72%**
Sensitivity	0.8378	0.8396	0.8470
Specificity	0.5078	0.7390	0.7952

The results of a similar analysis performed using EMG signals are tabulated in Tables 6 through 9. HOS and Non-linear of EMG signals resulted in a maximum accuracy of 75.5%. The values of sensitivity and specificity are specified in Table 9.

Though the accuracy of EMG signals is relatively less compared to ECG signals, certain emotions are better classified using EMG. Merging the features of ECG and EMG led to improved performance in identifying the individual emotional states. The accuracy of classification using Principal Component Analysis (PCA) is as tabulated in Table 10.

It can be inferred that the accuracy of identifying the six emotional states vary from 76.57% ti 92.47%. In most of the e-Learning systems, positive and negative emotions will suffice to enable productive learning. The obtained accuracy can be improved if two or three emotional states can be taken into consideration.

Table 6. Accuracy of EMG features after EMD

Feature	Classifier	Neutral (%)	Happiness (%)	Sadness (%)	Fear (%)	Surprise (%)	Disgust (%)	Average (%)
Statistical Features	Regression Tree	63.58	57.67	59.26	62.35	61.35	52.17	59.40
	Naïve Bayes	20.99	43.56	12.96	39.51	56.44	38.51	35.33
	KNN (K=8)	50.62	42.94	59.26	53.70	51.53	46.58	50.77
	FKNN (K=9)	52.47	45.40	58.02	50.62	53.99	45.96	51.08
HOS Features	Regression Tree	44.44	52.15	47.53	51.23	49.69	34.78	46.64
	Naïve Bayes	20.37	52.76	22.22	0.62	50.92	24.84	28.62
	KNN (K=6)	43.83	53.99	40.74	51.23	52.76	42.86	47.57
	FKNN (K=11)	41.98	53.37	51.85	53.70	58.90	42.86	50.44
Nonlinear Features	**Regression Tree**	**69.14**	**64.42**	**79.01**	**84.57**	**60.74**	**66.46**	**70.72**
	Naïve Bayes	56.17	71.17	78.40	70.37	56.44	81.99	69.09
	KNN (K=10)	56.79	68.10	70.37	78.40	53.37	62.11	64.86
	FKNN (K=8)	57.41	65.03	68.52	77.16	58.28	65.22	65.27

Table 7. Accuracy of EMG signal after EMD and HHT

Feature	Classifier	Neutral (%)	Happiness (%)	Sadness (%)	Fear (%)	Surprise (%)	Disgust (%)	Average (%)
Statistical Features	Regression Tree	65.43	60.12	64.20	61.11	60.74	54.04	60.94
	Naïve Bayes	41.36	39.26	33.95	61.11	33.13	19.25	38.01
	KNN (K=9)	52.47	51.53	50.00	58.02	47.24	43.48	50.46
	FKNN (K=11)	52.47	48.47	50.62	58.64	56.44	51.55	53.03
HOS Features	Regression Tree	44.44	52.15	52.47	45.06	52.76	43.48	48.39
	Naïve Bayes	23.46	58.28	12.35	42.59	38.04	27.95	33.78
	KNN (K=8)	54.94	52.15	47.53	46.91	66.87	42.86	51.88
	FKNN (K=10)	54.32	53.99	48.77	42.59	61.35	43.48	50.75
Nonlinear Features	**Regression Tree**	**59.26**	**69.94**	**80.86**	**79.63**	**79.75**	**83.85**	**75.55**
	Naïve Bayes	57.41	61.96	65.43	78.40	78.53	83.23	70.83
	KNN (K=10)	69.14	68.10	67.90	74.69	67.48	79.50	71.14
	FKNN (K=7)	59.26	58.90	59.26	69.75	64.42	72.67	64.04

Table 8. Accuracy of EMG signals after EMD and FFT

Feature	Classifier	Neutral (%)	Happiness (%)	Sadness (%)	Fear (%)	Surprise (%)	Disgust (%)	Average (%)
Statistical Features	Regression Tree	56.79	68.10	67.90	69.14	69.33	70.81	67.01
	Naïve Bayes	11.73	54.60	48.77	54.94	47.24	18.63	39.32
	KNN (K=10)	56.79	65.64	42.59	53.09	57.06	62.73	56.32
	FKNN (K=6)	51.23	60.74	46.30	57.41	59.51	70.19	57.56
HOS Features	Regression Tree	64.81	55.21	69.14	51.85	66.26	62.11	61.56
	Naïve Bayes	0.00	42.33	42.59	30.86	57.06	12.42	30.88
	KNN (K=12)	60.49	52.15	61.11	47.53	57.67	61.49	56.74
	FKNN (K=9)	51.85	50.31	58.02	49.38	58.90	65.22	55.61
Nonlinear Features	**Regression Tree**	**69.14**	**71.78**	**61.11**	**59.88**	**71.17**	**74.53**	**67.93**
	Naïve Bayes	28.40	61.96	46.30	50.62	80.37	70.81	56.41
	KNN (K=6)	41.98	56.44	49.38	48.15	49.69	39.75	47.57
	FKNN (K=7)	48.15	62.58	53.70	55.56	54.60	55.90	55.08

Table 9. Performance Metrics of EMG based Emotion Recognition

	EMD	HHT	FFT
Accuracy	70.72%	75.75%	67.93%
Sensitivity	0.9149	0.947	0.9093
Specificity	0.6515	0.7793	0.655

Table 10. PCA analysis of emotional features from ECG and EMG signals

Classifier	Neutral (%)	Happiness (%)	Sadness (%)	Fear (%)	Surprise (%)	Disgust (%)	Average (%)
Regression Tree	67.43	68.00	83.33	82.86	75.43	76.30	75.56
Naïve Bayes	40.00	77.14	63.22	83.43	61.71	79.19	67.45
KNN (K=6)	74.86	76.00	81.61	87.43	75.43	90.75	81.01
FNN (K=7)	**80.00**	**76.57**	**81.61**	**86.29**	**78.29**	**92.49**	**82.54**

3.2 Integration of Emotion Recognition into e-Learning System

The GUI developed to integrate emotion recognition in e-Learning system is as shown in Figure 11. The GUI consists of four main modules namely, the input module, the output module, the suggestion module and the menu for performing the suggestions. This GUI uses both ECG and EMG signals to find the emotional states. The different modules are explained below

Input module: The input module is used to provide input to the e-Learning System. In this developed module the emotional ECG and EMG data are obtained from the local device. Once the ECG and EMG signals are selected, appropriate pre-processing is done. The raw and pre-processed are plotted to understand if the noises are removed. Figure 11 portrays the navigation when 'Choose Emotional ECG signal' is clicked. Similarly 'Choose Emotional EMG signal' opens the folder that contain the EMG signals. When developing in real time, the sensors used to collect the ECG and EMG data need to be integrated in the Input module.

Output module: The Output module displays the emotion that is recognized by the underlying classification algorithm. The emotional features obtained from the signals obtained in the input module is fed into the developed emotion recognition model to identify the emotional state.

Suggestion module: Each emotional state has its own challenges and mechanisms of furtherance. This module can be personalized and provide suggestions for action, as needed by the user. In this case, the learner can continue learning by clicking on 'START LESSONS' menu when he/she is in the normal or positive emotional state. Music therapy is used to calm the learner when he is sad or fearful. The learner is asked to calm down by remaining silent for a few minutes, when in a surprise state. Similarly, during disgust, the minds are diverted by engaging in a computer game. The emotions can be checked by the user as and when needed. The suggestions need to be personalized to suit the user.

Menu module: This module contains the navigation system for the user towards his lessons. He can 'START LESSONS', 'LISTEN MUSIC' or 'PLAY A GAME' as suggested by the suggestion module. These navigation can also be automated and personalized in the e-learning system as needed by the user.

Figure 14 shows the emotions displayed and the suggestion specified to the learner. This software prototype works on offline mode and should be integrated with data acquisition sensors to work on an online mode.

FUTURE WORK

An exhaustive environment of emotion aware e-Learning systems need to take several other metrics into consideration. The developed e-Learning system works on offline data with an average accuracy of 82%. The accuracy of the system may be improved by including more data and trying newer hybrid algorithms. Behavioral measures may also be integrated into the algorithm to improve the efficiency. Emotions such as boredom and anxiety during learning may be taken into consideration when developing a protocol to acquire emotion data.

This algorithm also needs to be integrated to work with real time data, that could notify the emotion of the learner, as and when needed. Time duration of study is also an important factor and prompts may be provided to include active learning in the students.

REFERENCES

Alhazzani, N. (2020). MOOC's impact on higher education. *Social Sciences & Humanities Open*, *2*(1), 100030. doi:10.1016/j.ssaho.2020.100030 PMID:34171022

Bunluechokchai, C., & Leeudomwong, T. (2010). Discrete Wavelet Transform -based Baseline Wandering Removal for High Resolution Electrocardiogram. *Int J Applied Biomed Eng, 3*.

Chauhan, J., & Goel, A. (2017). An Overview of MOOC in India Education Management View project An Overview of MOOC in India. *International Journal of Computer Trends and Technology*, *49*(2), 111–120. Advance online publication. doi:10.14445/22312803/IJCTT-V49P117

Crane, R. A., & Comley, S. (2021). Influence of social learning on the completion rate of massive online open courses. *Education and Information Technologies*, *26*(2), 2285–2293. doi:10.100710639-020-10362-6

Haag, A., Goronzy, S., Schaich, P., & Williams, J. (2004). Emotion Recognition Using Bio-sensors: First Steps towards an Automatic System. *Affective Dialogue Systems*, *i*, 36–48. doi:10.1007/978-3-540-24842-2_4

Hamann, K., Glazier, R. A., Wilson, B. M., & Pollock, P. H. (2021). Online teaching, student success, and retention in political science courses. *European Political Science*, *20*(3), 427–439. doi:10.105741304-020-00282-x

Hökkä, P., Vähäsantanen, K., & Paloniemi, S. (2020). Emotions in Learning at Work: A Literature Review. *Vocations and Learning*, *13*(1), 1–25. doi:10.100712186-019-09226-z

Imani, M., & Montazer, G. A. (2019). A survey of emotion recognition methods with emphasis on E-Learning environments. *Journal of Network and Computer Applications*, *147*(April), 102423. doi:10.1016/j.jnca.2019.102423

Jerritta, S., Murugappan, M., Nagarajan, R., & Wan, K. (2011). Physiological signals based human emotion Recognition: A review. *Signal Processing and Its Applications (CSPA), 2011 IEEE 7th International Colloquium On,* 410–415. 10.1109/CSPA.2011.5759912

Min, H., & Nasir, M. K. M. (2020). Self-Regulated Learning In A Massive Open Online Course: A Review of Literature. *European Journal of Interactive Multimedia and Education*, *1*(2), e02007. doi:10.30935/ejimed/8403

Muñoz, S., Sánchez, E., & Iglesias, C. A. (2020). An emotion-aware learning analytics system based on semantic task automation. *Electronics (Switzerland)*, *9*(8), 1–24. doi:10.3390/electronics9081194

Picard, R. W., Vyzas, E., & Healey, J. (2001). Toward machine emotional intelligence: Analysis of affective\nphysiological state. *IEEE Transactions on Pattern Analysis and Machine Intelligence*, *23*(10), 1175–1191. doi:10.1109/34.954607

Selvaraj, J., Murugappan, M., Wan, K., & Yaacob, S. (2013). Classification of emotional states from electrocardiogram signals: A non-linear approach based on Hurst. *Biomedical Engineering Online*, *12*(1), 44. doi:10.1186/1475-925X-12-44 PMID:23680041

Shah, D. (2020). *By The Numbers: MOOCs in 2020*. The Report by Class Central. https://www.classcentral.com/report/mooc-stats-2020/

Shen, L., Wang, M., & Shen, R. (2009). Affective e-Learning: Using "emotional" data to improve learning in pervasive learning environment related work and the pervasive e-learning platform. *Journal of Educational Technology & Society*, *12*, 176–189.

Stephan, M., Markus, S., & Gläser-Zikuda, M. (2019). Students' achievement emotions and online learning in teacher education. *Frontiers in Education*, *4*(October), 1–12. doi:10.3389/feduc.2019.00109

Tyng, C. M., Amin, H. U., Saad, M. N. M., & Malik, A. S. (2017). The influences of emotion on learning and memory. *Frontiers in Psychology*, *8*(Aug), 1454. Advance online publication. doi:10.3389/fpsyg.2017.01454 PMID:28883804

Chapter 4
A Predictive Model Emotion Recognition on Deep Learning and Shallow Learning Techniques Using EEG Signal

Vidhya R.
SRM Institute of Science and Technology, India

Sandhia G. K.
SRM Institute of Science and Technology, India

Jansi K. R.
SRM Institute of Science and Technology, India

Nagadevi S.
SRM Institute of Science and Technology, India

Jeya R.
SRM Institute of Science and Technology, India

ABSTRACT

Social, psychological, and emotional well-being are all aspects of mental health. Mental illness can cause problems in daily life, physical health, and interpersonal connections. Severe changes in education, attitude, or emotional management of students cause suffering are defined as children's mental disorders. Artificial intelligence (AI) technology has lately been advanced to help intellectual fitness professionals, especially psychiatrists and clinicians, in making choices primarily based totally on affected person records along with medical history, behavioural records, social media use, and so on. There is a pressing need to address core

DOI: 10.4018/978-1-6684-3843-5.ch004

Copyright © 2023, IGI Global. Copying or distributing in print or electronic forms without written permission of IGI Global is prohibited.

mental health concerns in children, which can progress to more serious problems if not addressed early. As a result, a shallow learning technique-assisted integrated prediction model (SLIPM) has been presented in this research to predict and diagnose mental illness in children early. Convolutional neural networks (CNN) are built first in the proposed model to learn deep-learned patient behavioural data characteristics.

INTRODUCTION

Emotion popularity is a method for knowledge and extracting the modern human intellectual kingdom or modes of mind. Emotion is a critical issue of being human, and it has a good sized effect on each day sports like communication, interaction, and learning. The purpose of this studies is to broaden an EEG-primarily based totally emotion detection gadget which could inform the distinction among 3 distinctive emotions: positive, neutral, and negative(Fink, M 2017)(Bahari, F., and Janghorbani, A 2013). Up to this date, numerous modelling methods for computerized emotion popularity were documented. However, the temporal dependency belongings become now no longer absolutely investigated all through the emotion process. Furthermore, computerized emotion popularity is an critical and hard subject matter withinside the subject of human-system interaction (HMI). The development of Artificial Intelligence (AI) technology, emotion popularity has grow to be a important thing of studies withinside the domain names of neurology, pc science, cognitive science, and scientific science. Furthermore, emotion detection from speech, gesture, and posture turns into intricate for inarticulate or bodily challenged individuals who can not speak or explicit their sentiments via gesture or posture. As a result, EEG is a viable method for extracting human emotion and has already been used in numerous investigations to analyse human emotion. Nowadays, machines, particularly robots, are used in a wide range of industries, hospitals, and even domestic applications. As robots grow increasingly widespread in many aspects of daily life, people are establishing higher expectations for them. The super ability of decision making, self-thinking, and emotion detecting is hoped for to improve human-machine interaction. Emotion recognition assurance is an unavoidable requirement for making a robot more practical for real-world applications. The patient's affective information, which includes his or her emotional state, is a critical aspect in determining his or her mental and physical well-being. The emotional state of a patient has a substantial impact on treatment management.

Many signals have been authorised, adopted, and roughly divided into non-physiological and physiological signals in the practical application of emotion recognition. Speech, gesture, facial expression, movement, voice intonation, and text, among other non-physiological signals, are largely utilised in earlier work. More study has recently been conducted using physiological signals such as EEG, electrocardiogram (ECG), pupillary diameter (PD), and electromyogram (EMG), all of which are more effective and dependable. EEG uses electrodes on the scalp to record brain activity in the central nervous system and provides useful information about emotional responses. Neurologists use the EEG signal to analyse and diagnose a variety of brain problems, including seizure detection, autism, attention deficit, and game addiction.

Emotion is a concept, a feeling, or a conscious experience in which people are exposed to internal or external stimuli. Emotion plays a crucial part in natural communication between humans and other living things. EEG indicators were extensively used to increase green mind pc interplay structures for evaluation of each inner emotion and cognitive states. Emotion popularity primarily based totally on a unmarried modality, the EEG indicators were extensively used to increase green mind pc interplay structures for evaluation of each inner emotion and cognitive states. The aim of this paper is to apply a deep convolutional neural network (CNN) with a residual neural network (ResNet50) and the Adam optimizer to recognize positive, neutral, and terrible emotion the usage of the EEG dataset (SEED) primarily based totally on deep studying and shallow studying techniques. The following sections make up the shape of the paper: section I provides background information on Emotion Recognition and the Cognitive System. Section II: A synopsis of related work Experiments in Section III .Section IV is dedicated to methodology, Section V is dedicated to results and discussion, and Section VI is dedicated to conclusions and future study.

RELATED WORK

On the SEED Dataset, Santamaria-Granados et al. used a deep convolutional neural network. To get the features of physiological signals in the time, frequency, and nonlinear fields, the researchers applied advanced standard machine learning algorithms. The classification of emotional states is more precise with this method. After making use of long short-time period memory (LSTM) to stumble on functions from EEG indicators near the thick layer, Alhagry et al. utilised a deep gaining knowledge of technique to categorise emotion from uncooked EEG indicators, and functions had been classified into low/excessive arousal, valence, and liking successively. The DEAP dataset became utilised to validate this technique, which yielded common accuracy of 85.65%, 85.45%, and 87.99% for the arousal, valence, and liking classes, respectively.(Chi, Y. M., Wang, Y.-T., Wang, Y., Maier, C., Jung, T.-P., and Cauwenbe, G. (2012).)(Fink, M (2013))

The following eventualities will gain from this evaluate article: 1) The basics of emotion, EEG sign analysis, required software, to be had datasets, famous functions, and classifiers are addressed in detail, imparting new researchers with crucial concern knowledge. 2) A thorough overall performance evaluation primarily based totally on functions of deep getting to know and shallow device getting to know-primarily based totally class algorithms is provided, which might also additionally help intermediate-stage researchers in figuring out ahead studies areas. 3) A precis of some of very applicable articles is likewise included, together with their obstacles and recommendations, which might also additionally gain expert-stage researchers of their quest to expand the choicest emotion popularity device for real-world applications. The distinct lobes of the human brain and EEG sub-bands have a relationship with mental state and activity.(Chi, Y. M., Wang, Y.-T., Wang, Y., Maier, C., Jung, T.-P., and Cauwenbe, G. (2012))(Fink, M(2013))(Bohgaki, T., Katagiri, Y., and Usami, M. (2014).)

The classification of emotion recognition systems in general.

Figure 1. Classification of Emotion Recognition

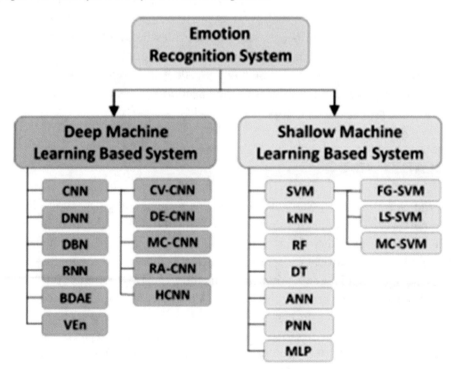

ELECTROENCEPHALOGRAM (EEG)

The electroencephalogram (EEG)(Cheng, B., and Liu, G(2013).) is a waveform recording era that data human mind electric pastime from the scalp over time. It video display units the voltage variation (withinside the microvolt range) as a result of the ionic contemporary flowing thru the mind's neurons. The human mind shape ought to be understood that allows you to reap correct and significant EEG data. (A., Shimohara, K., and Tokunaga, Y. (1989).) The mind, that is the centre of the Central Nervous System (CNS), is made from 3 parts: the cerebrum, cerebellum, and brainstem. The cerebrum, that is made from the proper and left hemispheres, is the most important of the 3. The frontal, parietal, temporal, and occipital lobes of the cerebrum, on the opposite hand, are separated into 4 lobes. The 5 subbands of EEG alerts are delta, theta, alpha, beta, and gamma, every of that is related with special intellectual states and situations.(Agrafioti, F., Hatzinakos, D., and Anderson, A. K. (2012)) Each sub-call band's is displayed, in addition to its location, frequency range, and mind pastime. The following are two methods for extracting human emotion:

(i)   using stimuli such as photos, audio, video, audiovisual, tactile, odour, and so on; and

(ii)   Asking people approximately any preceding emotional nation or existence situation. The first approach is now utilized by the bulk of researchers. Within the primary approach, about 26% used pictures as stimuli, 23.eight percentage used video, 17.five percentage used audio, and 22.2 percentage used an present dataset made of physiological and emotional statistics. The closing 10.five percentage of tasks took benefit of emotional statistics thru emotional games, stay performances, or existence events.

METHODS

Extraction and display of data was collected in the following categories:

A.   Task information • Task type • Number of subjects • Total duration of tested data

B.   Artifact elimination approach • Manual • Automatic • No cleansing or elimination

C.   Frequency variety that become used withinside the analysis

D.   Formulation of the input • Characteristics of EEG signals • Methods for channel selection(Hjorth, B. (1970))(Kübler, A., Furdea, A., Halder, S., Hammer, E. M., Nijboer, F., and Kotchoubey, B. (2009))

Tasks Specific Deep learning trends

In the responsibilities of emotion recognition, motor imagery, and sleep degree scoring, there has been no settlement on which deep getting to know algorithms to use. Research on seizure detection had been cut up nearly calmly among CNNs and RNNs, with the best percent of research utilizing RNNs in comparison to different responsibilities.(Abadi, M., Barham, P., Chen, J., Chen, Z., Davis, A., Dean, J., et al. (2016)) Only any such investigations used an SAE or MLPNN, and not one of the seizure detection experiments used DBNs. When in comparison to analyze utilizing CNNs, sleep degree scoring responsibilities had the best percent of research the use of hybrid formulations, which had been calmly represented. CNN's overall performance became definitely advanced in ERP research (the best percent of CNN research in comparison to all different responsibilities). The following strategies may be used to extract human emotion.(Fink, M (2013))(Bohgaki, T., Katagiri, Y., and Usami, M. (2014).)

DISCUSSION AND CONCLUSION

There are some factors which are really well worth debating. First, in contrast to preceding work, the counseled version can be educated using give up-to-give up manner. The give up-to-give up education approach became placed to the test, and it yielded comparable effects. However, through embedding an SAE shape right into a CNN (Fu, Q., Luo, Y., Liu, J., Bi, J., Qiu, S., Cao, Y., et al. (2017).) (Kira, K., and Rendell, L. A. (1992)). The education version can be in addition studied; the accuracy is higher than a CNN with the equal parameters and shape because the counseled community. Three separate capabilities are extracted for classifications withinside the proposed community. The proposed community's common popularity accuracy might also additionally reap 89.forty nine percentage on valence and 92.86 percentage on arousal for DEAP and 96.seventy seven percentage for SEED, wherein the counseled community has a quicker convergence speed, using PCC-primarily based totally characteristics. Furthermore, the period of the overlap has an effect on performance, with effects from the SEED dataset (Koelstra, S., Muhl, C., Soleymani, M., Lee, J.-S., Yazdan, A., Ebrahimi, T., et al(2015)) indicating that facts of eight seconds with a four 2nd overlap produces the nice effects.(Huang, Y.-J., Wu, C.-Y., Wong, A. M.-K., and Lin, B.-S(2015)) The facts processed through the SAE is likewise observed to be without difficulty labeled withinside the proposed community, indicating that the SAE is powerful at extracting capabilities from EEG facts. In destiny research, the SAE and different classifiers can be used to growth type performance.

REFERENCES

. A., Shimohara, K., and Tokunaga, Y. (1989). "EMG pattern analysis and classification by neural network," in Conference Proceedings, IEEE International Conference on Systems, Man and Cybernetics (Cambridge, MA: IEEE), 1113–1115. doi:10.1109/ICSMC.1989.71472

Abadi, M., Barham, P., Chen, J., Chen, Z., Davis, A., Dean, J., & ... . (2016). "Tensorflow: a system for large-scale machine learning," in *12th USENIX Symposium on Operating Systems Design and Implementation* (Savannah, GA: USENIX Association), 265–283.

Agrafioti, F., Hatzinakos, D., & Anderson, A. K. (2012). ECG pattern analysis for emotion detection. *IEEE Transactions on Affective Computing*, *3*(1), 102–115. doi:10.1109/T-AFFC.2011.28

Bahari, F., & Janghorbani, A. (2013). "EEG-based emotion recognition using recurrence plot analysis and k nearest neighbor classifier," in 2013 20th Iranian Conference on Biomedical Engineering (ICBME) (Tehran: IEEE), 228–233. 10.1109/ICBME.2013.6782224

Bohgaki, T., Katagiri, Y., & Usami, M. (2014). Pain-relief effects of aroma touch therapy with citrus junos oil evaluated by quantitative EEG occipital alpha-2 rhythm powers. *Journal of Behavioral and Brain Science*, *4*(01), 11–22. doi:10.4236/jbbs.2014.41002

Cheng, B., & Liu, G. (2008). "Emotion recognition from surface EMG signal using wavelet transform and neural network," in 2008 2nd International Conference on Bioinformatics and Biomedical Engineering (Shanghai: IEEE), 1363–1366. 10.1109/ICBBE.2008.670

Chi, Y. M., Wang, Y.-T., Wang, Y., Maier, C., Jung, T.-P., & Cauwenbe, G. (2012). Dry and noncontact EEG sensors for mobile brain– computer interfaces. *IEEE Transactions on Neural Systems and Rehabilitation Engineering*, *20*(2), 228–235. doi:10.1109/TNSRE.2011.2174652 PMID:22180514

Danelljan, M., Robinson, A., Khan, F. S., & Felsberg, M. (2016). "Beyond correlation filters: learning continuous convolution operators for visual tracking," in *European Conference on Computer Vision*, eds B. Leibe, J. Matas, N. Sebe, & M. Welling (Amsterdam: Springer), 472–488. 10.1007/978-3-319-46454-1_29

Fink, M. (1969). EEG and human psychopharmacology. IEEE Trans. Inform. Technol. Biomed. 9, 241–258. , Q., Luo, Y., Liu, J., Bi, J., Qiu, S., Cao, Y., et al. (2017). "Improving learning algorithm performance for spiking neural networks," in 2017 IEEE 17th International Conference on Communication Technology (ICCT) (Chengdu: IEEE), 1916–1919. doi: 10.1146/annurev.pa.09.040169.001325

García, H. F., Álvarez, M. A., & Orozco, Á. A. (2016). "Gaussian process dynamical models for multimodal affect recognition," in 2016 38th Annual International Conference of the IEEE Engineering in Medicine and Biology Society (EMBC) (Orlando, FL: IEEE), 850–853. 10.1109/EMBC.2016.7590834

Girshick, R., Donahue, J., Darrell, T., & Malik, J. (2014). "Rich feature hierarchies for accurate object detection and semantic segmentation," in 2014 IEEE Conference on Computer Vision and Pattern Recognition (Washington: IEEE), 580–587. Hiraiwa10.1109/CVPR.2014.81

Hjorth, B. (1970). EEG analysis based on time domain properties. *Electroencephalography and Clinical Neurophysiology*, *29*(3), 306–310. doi:10.1016/0013-4694(70)90143-4 PMID:4195653

Huang, Y.-J., Wu, C.-Y., Wong, A. M.-K., & Lin, B.-S. (2015). Novel active combshaped dry electrode for EEG measurement in hairy site. *IEEE Transactions on Biomedical Engineering*, *62*(1), 256–263. doi:10.1109/TBME.2014.2347318 PMID:25137719

. Kira, K., and Rendell, L. A. (1992). "A practical approach to feature selection," in Machine Learning Proceedings (San Francisco, CA), 249–256. doi:10.1016/B978-1-55860-247-2.50037-1

Koelstra, S., Muhl, C., Soleymani, M., Lee, J.-S., Yazdan, A., Ebrahimi, T., & ... . (2012). Deap: A database for emotion analysis; using physiological signals. *IEEE Transactions on Affective Computing*, *3*(1), 18–31. doi:10.1109/T-AFFC.2011.15

Kübler, A., Furdea, A., Halder, S., Hammer, E. M., Nijboer, F., & Kotchoubey, B. (2009). A brain–computer interface controlled auditory event-related potential (p300) spelling system for locked-in patients. *Annals of the New York Academy of Sciences*, *1157*(1), 90–100. doi:10.1111/j.1749-6632.2008.04122.x PMID:19351359

Chapter 5
Enhanced BiLSTM Model for EEG Emotional Data Analysis

Shanthalakshmi Revathy J.

iD https://orcid.org/0000-0003-1724-7117
Velammal College of Engineering and Technology, India

Uma Maheswari N.
PSNA College of Engineering and Technology, India

Sasikala S.

iD https://orcid.org/0000-0001-5972-6349
Velammal College of Engineering and Technology, India

ABSTRACT

Emotion recognition based on biological signals from the brain necessitates sophisticated signal processing and feature extraction techniques. The major purpose of this research is to use the enhanced BiLSTM (E-BiLSTM) approach to improve the effectiveness of emotion identification utilizing brain signals. The approach detects brain activity that has distinct characteristics that vary from person to person. This experiment uses an emotional EEG dataset that is publicly available on Kaggle. The data was collected using an EEG headband with four sensors (AF7, AF8, TP9, TP10), and three possible states were identified, including neutral, positive, and negative, based on cognitive behavioral studies. A big dataset is generated using statistical brainwave extraction of alpha, beta, theta, delta, and gamma, which is then scaled down to smaller datasets using the PCA feature selection technique. Overall accuracy was around 98.12%, which is higher than the present state of the art.

DOI: 10.4018/978-1-6684-3843-5.ch005

Copyright © 2023, IGI Global. Copying or distributing in print or electronic forms without written permission of IGI Global is prohibited.

1. INTRODUCTION

Human emotions are vital in daily life and influence daily activities (Bos, 2006). The goal of affective computing is to create an emotional model that can monitor and interface with human emotional states. The subjective nature of a person's inner emotions is based on feelings, and experiences, both internal and external to the individual (Alarcao & Fonseca, 2019). To name a few, voice, facial, and physiological signals can all be used to detect and evaluate emotional states. The subject might disregard or falsify their mood states, which can lead to erroneous choices, according to the defects of the speech and facial approach. These shortcomings have been overcome by analyzing using physiological signals (Liu et al., 2010). The use of electroencephalogram (EEG) to detect emotions is quickly rising due to factors such as no interference with brain signals and the availability of numerous portable data-gathering equipment. This has allowed for the development of medicinal and non-medicinal applications (Molina et al., 2009).

The importance of EEG has been vastly exaggerated over time, and it is far from a panacea for brain activity. It can be detected if someone is awake, asleep, brain dead, suffering a seizure, and a few other things clinically. The EEG is the sum of all electrical stimulation on the surface of the brain. Because this action must pass through layers of soft tissue, bone, and skin, the data is naturally noisy.EEG data is collected using a standard setup of 20 electrodes spread across the scalp. The letter in each lead denotes which section of the brain it is closest to (Temporal, Frontal, Parietal, and so on), with odd numbers and even numbers on the left and right respectively. In the clinic, usually consider the potential difference between pairs of electrodes rather than the electrical potentials at each electrode. This allows deducing what the brain is doing in that location by looking at the electrical field in the brain region between these two places. When any two electrodes are chosen and it generates 20 factorial distinct potential differences, not all of them will be beneficial.

As Montages, the arrangement of selecting pairs of electrodes to compare potential differences. There are several other montage systems, but the 10-20 system is the most prevalent. Looking at the firing rate is where the EEG data gets fascinating. The neuronal activity begins to synchronize in quite amazing ways with specific medical illnesses and mental states. This activity's firing rate is measured in Hz and divided into bands:

- Delta (<4Hz) Continuous attention activities, slow-wave sleep
- Theta (4-7Hz) Repression of evoked responses, drowsiness
- Alpha (8-15Hz) Closed eyelids, relaxed
- Beta (16-31Hz) Active thinking, concentration, and vigilance

- Gamma (>32Hz) Cross-sensory perception, short-term memory

So, if it's in Alpha or lower, they're either enjoying or falling asleep throughout the mission. In the case of beta or above, they're having problems concentrating on their distraction. The chapter is divided into sections. Section 2 focuses on relevant studies in this field. In section 3, the suggested work's methodology is outlined step by step. Section 4 discusses the observations and discussions, while Section 5 concludes.

2. RELEVANT STUDIES

Traditional machine learning and deep learning are used to develop EEG-based emotion identification systems. In classic machine learning-based emotion detection systems, features are manually acquired and entered into Naive Bayes, Support Vector Machine, and other classifiers for classification and identification. Based on the four physiological signals, a database with multi-modal physiological emotions was presented (Song et al., 2019). The PSD, FFT, and Hjorth parameters were used to extract features, while the classifiers used were k-Nearest Neighbor, support vector machine, and attention-long memory. (Shu et al., 2018) described a typical machine learning technique for EEG emotion detection that includes emotion formation, signal gathering, extraction of features, and classifier identification. Simultaneously, the flaws in standard machine learning methods were exposed, revealing the future path of EEG emotion recognition. Because EEG signals are non-linear and high-dimensional, using a linear method to identify them is difficult.

(Sharma et al., 2020) achieves 82.01 accuracy in live identification of human emotions based on electroencephalogram (EEG) inputs using an SVM classifier with RBF kernels. The mean-standard deviation limit can be used to decide whether to discriminate between high and low arousal and valence emotions. LSTM attention and the LSTM attention CNN model achieved 2 and 3 level classifications in (Kim & Choi, 2020). The two-level and three-level classifications of the LSTM and attention models did relatively well in the high class, whereas the LSTM attention CNN model performed better in the medium class. This proved that using CNNs to extract features is advantageous for EEG signal processing. The DEAP-Database is used as an input for ANN to classify specific emotions like Angry, Happy, Sad, and Relax in (Hemanth, 2020). For feature detection, time domain and time-frequency domain are used, and Support Vector Machine is used for classification, yielding an accuracy of 87.27 percent. To increase the accuracy of the classification of human emotions. Modified Kohonen neural networks I and II achieve a 1-2 percent improvement. (Phan et al., 2021) Feature-homogeneous matrices were used to organize the time domain. For feature extraction, we employed time, frequency, time frequency, and correlation, and for classification, we used multi-scale kernel CNN. It obtains arousal accuracy of 98.27 and valence accuracy of 98.36. The best sort of features for our technique was time-domain based features, and channel and band correlation qualities increased EEG-based emotion identification performance. (Khateeb et al., 2021) SVM is used to obtain wavelet, time, and frequency (α, β, γ) characteristics. The entropy feature outperformed the other features in the wavelet domain, with an average accuracy of 65.19 percent for three frequency sub-bands (α, β, γ). Using a combination of wavelet entropy and time domain Hjorth parameters, the accuracy was increased to 65.92 percent. Deep learning allows for end-to-end mapping, which is crucial for dealing with non-linear problems. Deep learning-based emotion detection uses CNN, RNN, and LSTM frameworks to learn deep features and discriminate emotions automatically, substantially improving feature extraction. (Acharya et al., 2018) anticipated a deep learning framework consisting of a CNN with 13 layers. In this implementation, they have used 30 controls (15 normal and 15 depressed). The accuracy for the left and right hemispheres is 93.5% and 96% respectively. A most similar study was conducted by (Ay et al., 2019) having a deep neural network model integrating CNN-LSTM with a right hemisphere accuracy of 97.66 percent and a left hemisphere accuracy of 99.12 percent. Deep learning architecture is developed utilizing a one-dimensional CNN with LSTM (1DCNN-LSTM) model in (Mumtaz & Qayyum, 2019). This study demonstrated 95.97 percent classification accuracy, 99.23 percent precision, 93.67 percent recall, and 95.14 percent f-score. (Li et al., 2019) reported R2G-STNN, which comprises neural network models with the temporal and spatial region to the global hierarchical feature learning process, to integrate discriminative spatial-temporal EEG information. The modified BiLSTM network is utilized to search for spatial characteristics in this study.

3. METHODOLOGY

For successful emotion prediction, a hybrid deep learning (DL) model based on a CNN and bi-direction LSTM (BiLSTM) is proposed in this chapter. Other recently developed DL models are compared to the proposed hybrid DL model. The results suggest that the proposed hybrid DL model is reliable, outperforming previously reported DL models in terms of emotion prediction. Video snippets are used to create emotional moods in the proposed work. E-BiLSTM framework is used to classify the retrieved temporal frequency and statistical information into three (3) category states (Positive, Negative, and Neutral).

Figure 1. Depicts the methods of the proposed system.

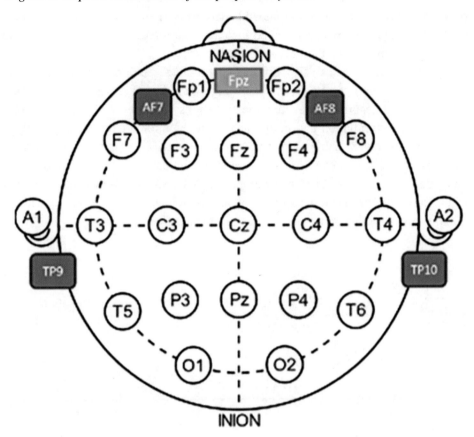

Protocol for Data Acquisition: The study makes use of four dry extracranial electrodes that are attached to a commercially available MUSE EEG headgear. Figure 1 shows micro voltage observations on the electrodes TP9, TP10, AF7, and AF8. Two volunteers (1 male, and 1 female, aged 20-22) collected 60 seconds of data for each of the six video clips, around 12 minutes of EEG signal data which carries the emotional state. Each emotional state has 6 minutes. The participants' EEGs were recorded for 36 minutes, including 6 minutes of neutral EEG signal data. This yielded a total of 324,000 data points derived from brain waves of various frequencies resampled to 150Hz. As the initial stage in normalizing the dataset, a Fourier-based technique was utilized to resample the data to a fixed frequency of 200Hz. Through a sliding window of 1second commencing at t=0 and t=0.5, statistical features such as Mean, Standard Deviation, Moments, Maximum value, Minimum value, Covariance matrix, Eigenvalue, Entropy, and correlation values were extracted. The minimum detected frequency of 150Hz was used for downsampling. The dataset was reduced to 2,549 source characteristics using feature selection techniques. The resulting dataset has 2549 columns and 2132 rows. The normalized mean value of the AF7 electrode, which revealed the classification of positive, negative, and neutral, was shown to be the most relevant feature.

Activities were exclusively stimuli that produced emotional responses from the emotions mentioned in Table I, and they were classified based on the labels of positive and negative rather than the sensations themselves. Neutral data were taken without stimulation and before any of the emotional data for a third class, which would show the subject's resting emotional state. Three minutes of data were gathered each day to lessen the influence of a calm emotional state. To avoid the data being impacted by Electromyographic (EMG) signals, which have a higher signal intensity than brainwaves, participants were told to observe the movie without making any conscious movements (e.g., drinking coffee).

Figure 2. EEG sensors TP9, AF7, AF8, and TP10

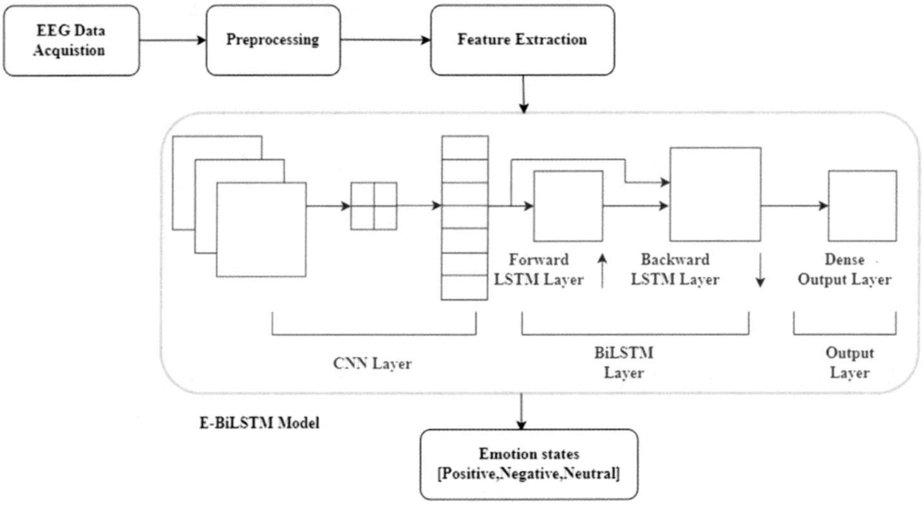

Preprocessing: Artifacts like muscle movements and eye blinking are present in the raw EEG data. The device featured a noise-canceling notch filter built in. There is also a median filter, a moving average referencing filter, and a 20th order bandpass FIR filter.

Feature Extraction: Principal component analysis (PCA) is the most widely utilized technique for pattern recognition and feature extraction. It reduces the number of variables. When there are a high number of variables and some duplication in the variables, PCA is utilized. Some of the variables are connected, which is known as redundancy. Because of this duplication, the observed variables may be reduced to a smaller number of primary components that account for the majority of the variation in the observed variables. A linear combination of ideally weighted observed variables can be characterized as the main component. PCA (Principal Component Analysis) is a technique for evaluating data and identifying patterns. PCA is a popular data reduction technique that converts higher-dimensional data into lower-dimensional data. When employing Principal Component Analysis with E-BiLSTM, the redundant data in the dataset is first removed, and the data is then trained with E-BiLSTM. The desired band of frequencies is extracted using 8-Level DWT decomposition on the preprocessed input. For each subject, a total of 34 features are extracted. Python is used to compute the features and extract channel-wise features extraction.

Classification: On the retrieved features, the Enhanced-BiLSTM (E-BiLSTM) classifier is used. To improve the prediction rate, a deep learning network was created by employing BiLSTM and a modified features vector framework.

Figure 3. Proposed E-BiLSTM Model

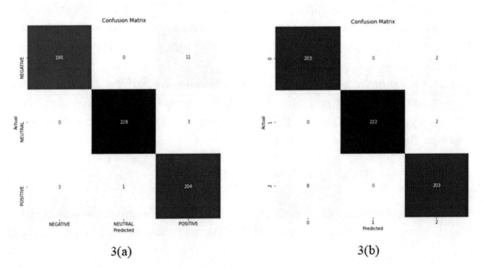

3(a)   3(b)

Figure 2 shows the proposed E-BiLSTM model in which CNN is combined with the BiLSTM model. The proposed model is divided into two stages. The first stage uses important information from EEG data to properly calibrate the emotional state. CNN is made up of three levels. The convolution layer is the first layer of CNN, and it executes convolution operations depending on kernel weight and EEG input data. The kernel, weight, and bias are all replicated in the relevant layers. The primary purpose for considering weight is to improve efficiency. ReLu is the action function employed here, which adds nonlinearity and executes the element-wise option. The activation function successfully handles RNN difficulties such as gradient disappearing and exploding. The second layer of CNN is the max-pooling layer, which receives input from the convoluted layer. The basic goal of the max-pooling layer is to extract as many features as possible. The data is converted to a 1-dimensional array by the flattening layer and then input into the proposed model's second stage.

The future state in LSTM is determined by the current and previous states. However, with BiLSTM, both past and future values are integrated and the present state is given more precedence to forecast the current state.

4. RESULTS AND DISCUSSION

The algorithm was implemented using Python, the Google TensorFlow numerical computation toolkit, and the Scikit-Learn machine learning package, as well as Jupyter Notebook as an IDE. The variation of the LSTM approach was used in this study. To calculate the accuracy of the LSTM, BiLSTM and E-BiLSTM models were compared with the actual and predicted values. Assessment indicators are important in the development of a model because they provide a proper prediction of emotion that requires improvement.

Accuracy is the most obvious criterion for evaluating performance. It's just the proportion of accurately predicted total observations. The E-BiLSTM obtained a 0.98 score, meaning that the model correctly predicted emotional state 98.12% of the time. The accuracy is given in Equation (1).

$$Accuracy = TP + TN \div TP + FP + FN + TN \tag{1}$$

The fraction of successfully predicted positive measurements to the total positive analyses expected is known as precision. The Precision measure is given in Equation (2).

$$Precision = TP \div TP + FP \tag{2}$$

Yes, remember the percentage of successfully predicted positive results to all actual class observations. The Recall calculation is given in Equation (3).

$$Recall = TP \div TP + FN \tag{3}$$

The F1-Ranking is a weighted ranking that is exact and easy to memorize. All false positives and false negatives are factored into the score. F1 is less straightforward to understand than accuracy, but it is usually more beneficial than accuracy, especially if the class distribution is inconsistent. Equation (4) shows the measure of the F1 Score.

$$F1\ Score = 2 * (Recall * Precision) \div (Recall * Precision) \tag{4}$$

The confusion matrix of the BiLSTM and E-BiLSTM models is shown in Figure (3). The performance of the E-BiLSTM model is superior to the BiLSTM model, as shown by the confusion matrix.

Figure 4: BiLSTM and E-BiLSTM Model Training, Validation Accuracy, and Loss

Figure 4. Confusion Matrix of BiLSTM and E-BiLSTM Model

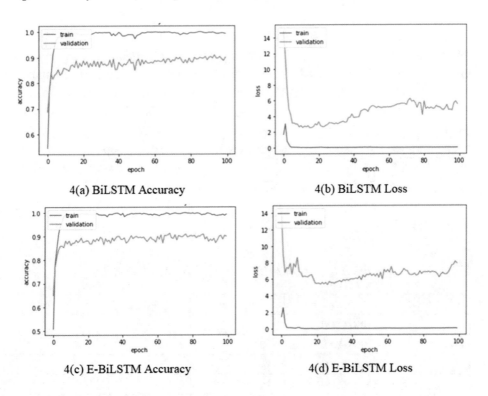

4(a) BiLSTM Accuracy   4(b) BiLSTM Loss

4(c) E-BiLSTM Accuracy   4(d) E-BiLSTM Loss

Figure 4 exhibits training, validation, and loss accuracy. Figure 4(a,b) depicts the BiLSTM model's training accuracy and loss across 100 epochs. Similarly, the E-BiLSTM model's training accuracy and loss are shown in four distinct ways (c,d). Validation accuracy or loss should be equal to or less than training accuracy or loss to avoid overfitting. According to the training and validation curves, the LSTM and BiLSTM models generalize well and do not overfit testing and validation data. After 40 epochs, the validation and training loss for the E-BiLSTM network did not provide a better curve, and the accuracy at the validation set decreased. It signifies that after 40 epochs, this network was unable to generalize and requires more parameters for future customization.

Table 1. Training, Validation Accuracy & Loss of BiLSTM and E-BiLSTM Model

Model	Accuracy	Validation Accuracy	Loss	Validation Loss
BiLSTM	0.9891	0.9532	0.7517	5.6390
E-BiLSTM	1.0000	0.9812	2.1239e-07	0.9212

The Training, Validation Accuracy, and Loss of the BiLSTM and E-BiLSTM Models are shown in Table 1. When compared to the validation accuracy of the BiLSTM model, the suggested model has higher validation accuracy. In comparison to the BiLSTM model, E-BiLSTM has a much lower validation loss. The performance of E-BiLSTM is now superior to that of other models. The classification report for the BiLSTM and E-BiLSTM Models is shown in Table 2.

As a consequence, substantial disparities between the two groups were discovered. The BiLSTM model's classification outcomes were accuracy = 95%, precision = 0.97, recall = 0.97, and f-score = 0.97. Furthermore, the study found that E-BiLSTM classification accuracy was 98.12%, precision was 0.98, the recall was 0.98, and f-score was 0.98.

Table 2. Classification Report of BiLSTM and E-BiLSTM Model.

Model	Emotion	Precision	Recall	F1-Score
BiLSTM	**NEGATIVE**	0.98	0.95	0.96
	NEUTRAL	1.00	0.99	0.99
	POSITIVE	0.94	0.98	0.96
E-BiLSTM	**NEGATIVE**	0.96	0.99	0.98
	NEUTRAL	1.00	0.99	1.00
	POSITIVE	0.98	0.96	0.97

This study's categorization findings were 98.12 percent accurate. The model can be further assessed with more study samples, which gives a more accurate picture of the deep learning architecture. Since 2015, the number of EEG-based machine learning (ML) algorithms for automated depression diagnosis has increased. A comparison of various studies is compared and the proposed model shows better accuracy shown in Table 3. In all the studies ensemble methods were used.

Table 3. A Comparison of this Study to Similar Works Performed on the same Datasets

Study	Method	Accuracy
(Bird, Ekart, et al., 2019)	InfoGain, RandomForest	97.89
(Bird, Ekart, et al., 2019)	InfoGain, MLP	94.89
(Bird, Ekart, et al., 2019)	Symmetrical Uncertainty, Ensemble Model	94.32
(Bird, Faria, et al., 2019)	AB(DEvo MLP)	96.23
(Bird, Faria, et al., 2019)	AB(LSTM)	97.06
This Study	BiLSTM	95.32
This Study	E-BiLSTM	98.12

The literature study shows over 85% of the studies use traditional machine learning methods and concluded positive research outcomes. On the other hand, just a few researchers have argued for the use of deep learning. The investigation also yielded useful classification results. The therapeutic implications of these technologies, on the other hand, are less obvious, and additional data on the efficacy of ML models is needed. This work adds to the body of knowledge by giving models. The study had a few limitations, such as the fact that antidepressant side effects should not be overlooked. More diverse study samples are required to test the universality of the proposed categorization models. As a result, extending the study's outcomes to other settings should be done with caution. Additionally, the proposed classification models should be generalized and evaluated on new study populations. Deep learning frameworks have the potential to revolutionize therapeutic applications for EEG-based depression detection. Based on the findings, it is reasonable to conclude that the deep learning framework might be used to automatically identify depression.

5. CONCLUSION

This chapter shows how deep learning architecture may be used to diagnose human emotions autonomously. The proposed deep learning framework does not require an explicit feature extraction stage, unlike other machine learning frameworks. Furthermore, the deep learning architecture for clinical situations has the potential to alter the way depression therapy is approached. Despite their flaws, classifier models have demonstrated their ability to distinguish between distinct emotional states. Finally, EEG-based deep learning might be used to predict human emotions in therapeutic contexts. In this work, an E-BiLSTM-based Deep Learning model for emotion prediction is developed. In terms of the loss function, the Deep Learning model based on E-BiLSTM outperforms other Deep Learning models in emotion prediction. These data show that the proposed DL model is a robust hybrid model capable of improving emotion prediction. The suggested model may be evaluated for both short and long-term emotion prediction in future research. Furthermore, hyperparameter adjustment can help the suggested DL model perform even better.

REFERENCES

Acharya, U. R., Oh, S. L., Hagiwara, Y., Tan, J. H., Adeli, H., & Subha, D. P. (2018). Automated EEG-based screening of depression using deep convolutional neural network. Computer Methods and Programs in Biomedicine, 161, 103–113. *doi:10.1016/j.cmpb.2018.04.012 PMID:29852953*

Alarcao, S. M., & Fonseca, M. J. (2019). Emotions Recognition Using EEG Signals: A Survey. IEEE Transactions on Affective Computing, 10(3), 374–393. *doi:10.1109/TAFFC.2017.2714671*

Ay, B., Yildirim, O., Talo, M., Baloglu, U. B., Aydin, G., Puthankattil, S. D., & Acharya, U. R. (2019). Automated Depression Detection Using Deep Representation and Sequence Learning with EEG Signals. Journal of Medical Systems, 43(7), 205. doi:10.1007109*16-019-1345-y PMID:31139932*

Bird, J. J., Ekart, A., & Faria, D. R. (2019). Mental Emotional Sentiment Classification with an E*EG-based Brain-machine Interface HANDLE Project (EU FP7) View project EMG-controlled 3D Printed Prosthetic Hand for Academia View project. Academic Press.*

*Bird, J. J., Faria, D. R., Mans*o, L. J., Ekárt, A., & Buckingham, C. D. (2019). A deep evolutionary approach to bioinspired classifier optimisation for brain-machine interaction. Complexity, 2019, 1–14. Advance online publicati*on. doi:10.1155/*2019/4316548

Bos, D. O. (2006). EEG-based emotion recognition - The Influence of Visual and Auditory Stimuli. Capita Selecta, 1–17.

Hemanth, D. J. (2020*). EEG signal* based Modified Kohonen Neural Networks for Classification of Human Mental Emotions. Journal of Artificial Intelligence and Sy*stems, 2(1), 1–13. doi:10.33969/AIS.2020.21001*

Khateeb, M., Anwar, S. M., & Alnowami, M. (2021). Multi-Domain Feature Fusion for Emotion Classification Using DEAP Dataset. IEEE Access: Practical Innovations, Ope*n Solutions, 9, 12134–12142. doi:10.1109/ACCESS.20*21.3051281

Kim, Y., & Choi, A. (2020). EEG-Based Emotion Classification Using Long Short-Term Memory Network with Attention Mechanism. Sensors (Basel), 20(23), 6727. doi:1*0.339020236727* PM*ID*:33255539

Li, Y., Zheng, W., Wang, L., Zong, Y., & Cui, Z. (2019). From Regional to Global Brain: A Novel Hierarchical Spatial-Temporal Neural Network Model for EEG Emotion Recognition. IEEE Transactions on Affective Co*mputing, 1–1. doi:10.1109/ TAFFC.2019.2922*912

Liu, Y., Sourina, O., & Nguyen, M. K. (2010). Real-Time EEG-Based Human Emotion Recognition and Visualization. 2010 International Conference *on Cyberworlds, 262–269. 10.1109/CW.2010.37*

Molina, G. G., Tsoneva, T., & Nijholt, A. (2009). Emotional brain-computer interfaces. 2009 3rd International Conf*erence on Affective Computing and Intelligent Interaction and Workshops, 1–9. 10.1109/ACII.2009.5349478*

Mumtaz, W., & Qayyum, A. (2019). A deep learning framework for automatic diagnosis of unipolar depression. International Journal of *Medical Informatics, 132, 103983. doi:10.10*16/*j.*ijmedinf.2019.103983 PMID:31586827

Phan, T.-D.-T., Kim, S.-H., Yang, H.-J., & Lee, G.-S. (2021). EEG-Based Emotion Recognition by Convolutional Neural Network with Multi-Scale Kernels. Sensors (Basel), 21(1*5), 5092. doi:1*0.339021155092 PMID:34372327

Sharma, R., Pachori, R. B., & Sircar, P. (2020). Automated emotion recognition based on higher order statistics and deep learning algorithm. Biomedical Signal *Processing and Control, 58, 101867. doi:10.10*16/j.bspc.2020.101867

Shu, L., Xie, J., Yang, M., Li, Z., Li, Z., Liao, D., Xu, X., & Yang, X. (2018). A Review of Emotion Recognition Using Physiological Signals. Sensors (Basel*), 18(7), 2074. do*i*:10.339018072074 PMID:29958457

Song, T., Zheng, W., Lu, C., Zong, Y., Zhang, X., & Cui, Z. (2019). MPED: A Multi-Modal Physiological Emotion Database for Discrete Emotion Recognition. IEEE Access: *Practical Innovations, Open Solutions, 7, 12177–12*191. doi:10.1109/ ACCESS.2019.2891579

Chapter 6
Harnessing the IoT– Based Activity Trackers and Sensors for Cognitive Assistance in COVID–19

Uma N. Dulhare
Muffakham Jah College of Engineering and Technology, India

Shaik Rasool
Methodist College of Engineering and Technology, India

ABSTRACT

The COVID-19 pandemic raised the need for harnessing digital infrastructure for many healthcare services like appointment scheduling, surveillance, and checking the patients remotely. A digital platform is needed that should be reliable for disease identification and monitoring using IoT, which can compensate for vital activities like the slow rate of viral tests and vaccine development also recognized as apt technology for bridging various devices. Although the technology has been used to connect the daily activities with the physical metrics, forecasting of COVID-19 is very vital and necessary. The fitness measures like body temperature, heartbeat rate, SPo2 from wearable devices can be used to alert the users. Groups of affected individuals can be remotely checked, and data can be collected to analyse the rate of transmission and symptoms. This chapter emphasizes harnessing the potential of IoT by comprehending the importance of IoT in various domains and its various applications. It will explore various IoT devices and focus on challenges and advantages and security aspects.

DOI: 10.4018/978-1-6684-3843-5.ch006

Copyright © 2023, IGI Global. Copying or distributing in print or electronic forms without written permission of IGI Global is prohibited.

1. BUDDING SIGNIFICANCE OF IOT IN THE MODERN ERA

Internet of Things (IoT) is assembly of heterogeneous devices that are equipped with sensors, software and various technologies that can connect, collaborate and interchange data. IoT devices can range from simple everyday objects in home to complex industrial applications. It is estimated by experts that currently there are around 10 billion IoT devices and the number may increase to 25 billion by the year 2025. IoT has grown into significant technology in this era. Embedded systems supported IoT to connect small appliances like thermostats, doorbells, baby monitoring systems to internet for storing and sharing data. Reliable communication was made possible through utilization of IoT among the devices and sensors which led to the application of IoT in multidisciplinary fields (Satyanarayana et al., 2022). IoT supported development of many modern technological innovations around the world which as not possible in past years. Various organizations and businesses benefitted from advancements made possible through IoT in supply, services and user experience.

IoT is also playing a significant role in harnessing the modern technologies like Artificial Intelligence, Machine Learning, Robotics and 5G for building application to sustain society. Complex tasks can now be executed with ease and enormous data generated from the IoT devices can be collected efficiently for analytics that can enhance the integration of AI and improve accuracy in automation with trivial errors. Past two years increased demand and need for IoT due to pandemic and lead to development of various innovative devices and applications. This demand will further increase in succeeding years and open doors for more research and opportunities. Imminent trends of IoT are discussed below that gives an insight into future stressing the need for IoT (Baidya & Levorato, 2018).

1.1 IoT Evolving as Apt Solution for Sustainable Innovations

Intelligent connected IoT devices are well capable of providing advanced applications which include power optimization, monitoring the environment and defence management are few to be named. Nevertheless, their importance in business was neglected by the major IoT developers till date. The shift of focus to green economy by several nations lead to increase in green initiatives for building smart cities and communities where IoT has huge potential and can provide the necessary techniques and infrastructure. Several cost-effective applications can be made possible that can be controlled and deployed remotely, effectively monitored for failures will increase extensively driving the IoT technology forward. IoT can be considered as a viable solution for global economy and should be integrated into suitable real-time application (Baidya & Levorato, 2018).

1.2 Transition to Edge Computing from Cloud Technology

Deep research in IoT platforms has gained traction from many developers in year 2015 as indicated by the IoT analytics. The current year has marked the beginning of race in industrial space for developing solution utilizing the potential of edge computing and would like themselves to be differentiated from the competition. Applications have become portable to be transitioned across different service providers by utilizing the virtualization techniques and containers. They relieve the business organization from being completely dependent on a specific vendor for services. It opens a wide variety of choice for the organisations to choose. These technologies have been adopted by most cloud service providers. Though cloud is offering several benefits it is not suitable for applications that need immediate processing of data and response. As edge computing involves processing of data at the edge i.e., near the data source, it may be essential when speed, bandwidth and scale are considering factors for building applications. As most advancements are based on IoT and that may require huge processing of data, putting that load on a central server like cloud may not yield best results (Simon, 2021). As transition from cloud to edge is happening at rapid pace it will benefit the IoT. For example, consider a self-driving car that needs immediate faster responses. If the processing is done at cloud, then it must wait few seconds to get response. Imagine there are few persons crossing the road and fell in front of car then a decision must be made in no time. Such critical situations require faster processing and instant response. Thus, edge computing will be only a viable solution for such problem.

1.3 Revolutionizing Manufacturing using Industrial IoT

IoT become most prominent and much needed technology in the current time of pandemic. Developers have often seen technology as a solution for every problem before even realizing the actual problem statement. IoT is designated to go beyond Supervisory Control and Data Acquisition (SCADA) for functioning of machines by making available all necessary information required for decision making about maintaining assets. IoT was looked like process of digitizing the data in previous years but now since smart analysis is available, manufactures can decide which problems can be solved on priority first.

1.4 Growth of Cloud Native Applications

Corporations now are making cloud adoption and migration a top precedence for the near time, as a minimum thru the next several years. Those which can be already closely invested inside the cloud are in search of new methods of riding performance and increasing talents, at the same time as the rest will need to rapidly increase migration plans. For a decade now, the cloud market has grown at a consistent fee, but the pandemic has considerably multiplied this growth, in phrases of both standard adoption and number of use cases. As cloud will become the norm for many at the infrastructure, platform, or software stage, the industry will see a whole new wave of programs advanced and optimized for cloud scale and performance, which in turn will assist supply improved reliability and reduce time to marketplace, meaning software program packages may be deployed quicker and with greater flexibly even as on the identical time lowering infrastructure complexities and charges (Wilford, 2022).

1.5 Hyper-Automation is Renovating Operations

Cutting-edge conventional awareness is that AI is the important thing to remodelling any organization utility, but the truth is that presently, most AI needs critical "facts ditch digging" to get to the point of assisting an agency. AI is simplest part of the transformational equation, and the second missing element is robotic system automation or RPA. And while AI and RPA are efficaciously joint and carried out, the result is hyper-automation. The pandemic created an inflection factor, prioritizing workers' safety and the technologies needed to aid them, and the labour shortage which began before the pandemic has end up even more tough of a constraint to cope with, that is accelerating using hyper-automation to enhance manner performance from the store ground to the top floor.

1.6 AI is Being Adopted Highly at the Edge

Corporations have been rethinking wherein to location their AI workloads, inside the cloud or at the threshold. So far, AI area applications have run on compute- and electricity-intensive side gadgets like business computers and part routers. However, two developments are riding a shift to the thin facet. Developments in semiconductors mainly on the decrease-fee, decrease-power cease of the spectrum, imply that AI can be placed ever toward the smallest stage of device. There is each purpose to agree with many microcontrollers (MCU's) will have on-tool AI within the near time period. Ai algorithms have become greener inside the closing five years. For instance, current ai algorithms need for much less compute strength than only a few years in the past to teach a neural community for visual item popularity. Some experts have postulated a 2x reduction in compute strength wished each sixteen months.

1.7 Adoption of Invisible AI

the purpose of replicating human cognition has been mentioned and well-liked for thousands of years. Artificial intelligence isn't a product. You may argue it isn't even a generation. It isn't an invention both. There is no single day we will mark as its start and we most likely gained recognise when it ends. Yet it is already basically ubiquitous! Pushed with the aid of improvements in other technology, consisting of computing energy, value of computing sources, the net, sensors, and most of all the improvement of complicated gadget gaining knowledge of algorithms and fashions. More than something, AI is a quest; an adventure striving toward the goal of creating an intelligence, and as such it's far constantly improving, expanding, and converting. What makes AI unique from maximum other industry four. 0 technology is its potential to harness creativity and imagination. Ai is extremely flexible in its utility – it spans each enterprise and affects nearly every activity feature – from access-stage up to CEO. It has a mysterious, technology-fiction attraction that captivates humans' curiosity and creativeness. But, due to the fact many human beings conflict to apprehend precisely how it works, a part of the population is afraid of using AI to help with decision-making.

1.8 Adoption of VR/AR in Enterprise Environment

Even as human beings are already dwelling and operating in smart environments, locating more efficient and powerful ways of coexisting in these spaces has emerge as an essential consciousness for researchers and businesses. Which means growing operational efficiency and exceptional in running environments, but also developing faraway operation management. The covid-19 pandemic has bolstered the attention on the latter. For this, the convergence of IoT with immersive reality technologies and surroundings simulation technology (including virtual twins) is essential. Considering this sort of convergence would require quite a few records, the upward push of 5g will boost up the improvement of VR and AR programs for organization and business applications.

1.9 5G is Shaping for IoT

Purchaser-oriented pundit's concept of 5G as really a mechanism for quicker downloads. The real strength for business and industrial functions lies in personal networks constructed on 5G. By the cease of 2022, 5G will be considered simply another device. Engineers use it no longer thinking about being pioneers but pretty much fixing a trouble.

1.10 Secure Remote Access of Assets is Growing in Importance

Groups that have enabled remote to get right of entry to their machines stand to understand numerous advantages, together with better support from gadget providers that could remotely troubleshoot and examine system records in actual-time, in addition to access to remote challenge count experts and decrease-cost labour. However, with expanded get entry to come elevated cyber protection attack vectors, so groups are after increasing their spending on cybersecurity to address this. Three key use cases for cyber safety in 2022 are asset visibility, deep-packet inspection, and 0 consider architectures. Asset visibility software materials special records approximately all nodes and customers linked to a community, allowing directors to locate suspicious users or gadgets quickly. Deep-packet inspection can alert operators of industrial networks when suspicious activity is spotted inside the industrial protocols that run those networks.

2. IoT EVOLUTION DURING COVID-19

Throughout the beyond 12 months covid-19 has made the merits of IoT technologies clearer than ever before. The pandemic has highlighted the significance of commercial enterprise agility in dealing with huge-scale crises that reason modifications in commercial enterprise styles and clients' behaviours. Bodily interactions (e. G., bodily conferences) were reduced to a minimum and changed through new, virtualized approaches (Mazumder, 2021). Furthermore, establishments put the fitness and protection in their personnel at the very top in their agendas. On this context, IoT has performed a key function in enhancing company agility and ensuring business continuity in diverse enterprise sectors consisting of clever homes and centers management. Outstanding examples of IoT packages on this course include:

2.1 Vacancy Monitoring in Resources Administration

leveraging sensors and IoT analytics, establishments have controlled to check the occupancy of rooms and workplaces in buildings and different centers. The detected occupancy styles have been used to ideal the operation of HVAC (heating, ventilation and air conditioning) structures towards optimizing energy fees, environmental overall performance and the tenants' consolation. This will have been hardly viable without IoT, as covid-19 disrupted the usual occupancy patterns of operating spaces (Fossum Færevaag, 2021).

2.2 Remote Supervision of Assets

IoT devices have enabled enterprises to check and manage their assets remotely. This has obviated the need for physical on-site inspections. Moreover, it has boosted the implementation of intelligent asset management strategies (e.g., predictive maintenance) that perfect field service schedules and reduce assets' downtime.

2.3 Employees' Health Care Monitoring

IoT devices have enabled organizations to check and control their property remotely. This has obviated the want for bodily on-web page inspections. Furthermore, it has boosted the implementation of smart asset management techniques (e. G., predictive upkeep) that ideal subject service schedules and reduce assets' downtime.

2.4 Workplace Safety and Wellness

companies have leveraged IoT to examine occupancy styles towards evidence-based totally place of job optimization. As an instance, they've scrutinized employee's density to evaluate and optimize the to be had workspace for group protection, health and happiness. Moreover, they've derived insights into the effect of various occupancy patterns on their operational charges.

2.5 Smart Cleaning

COVID-19 has unveiled IoT's large potential for sensible cleaning of workplaces. IoT gadgets have been used to song hygiene tactics in public areas and private facilities. This has eased the implementation of clever cleaning practices (e. G., cleaning signals) and boosted compliance with hygiene policies.

2.6 Legionella Monitoring

every other fallout from covid-19 has been the prediction of a legionella outbreak, following the huge-scale abandonment of public spaces at some stage in lockdown. This has left pipework disused and open to turning into a probable breeding ground for pathogens. Pipe tracking for water waft and temperature range can now be completed remotely, reducing the want for on-web site interest and probably lowering waste and fee.

3. MAJOR APPLICATIONS OF IOT

The Internet of Things, combined with other technological disruptions like 5G, automation, and machine learning, revolutionize the way we live and do business. In that vein, let's look closer at IoT and some of the practical real-world IoT applications already making a positive impact today.

3.1 Smart Home

'Smart Home' is Google's most searched IoT feature at present, with IoT creating the buzz. However, what does the word "Smart Home" mean? How nice it would be if you could turn the air conditioning on before you reached home or turn off the lights after you left? Or while you are away, let friends into your home for temporary access.

Smart homes are creating a revolution in the residential space and are expected to become as common as smartphones. They are being touted as practical, efficient, and cost-effective. An example is smart thermostats that can check and control the temperature in a home according to the owner's comfort. Moreover, there is smart lighting, which can adjust itself according to user preferences as well as external ambient lighting (Kim et al., 2015).

3.2 Wearables

In the IoT world, wearable technology is mandate, and probably one among initial applications of IoT. We have now become accustomed to wearing wearable devices, for example, virtual glasses, fitness bands that measure heartbeats and calories, GPS tracking belts, and smartwatches, among others (Hurford et al., 2006). Essentially, these are energy-efficient, small and compact devices that contain sensors, hardware for readings, measurements, and software to collect and organize data about users. Today, wearable devices can display calls, texts, and social media updates in addition to tracking health and fitness.

3.3 Smart Cities

IoT applications such as smart cities generate a lot of buzz among the world's population. An optimized traffic system is one of the many aspects of a smart city. Smart cities use the internet of things in many ways, such as water management, smart energy management, smart surveillance, automated transportation, smart street lighting, waste management and environmental monitoring, etc.

Urban dwellers will find IoT helpful in solving major problems like pollution, traffic congestion, and energy shortages (Founoun & Hayar, 2018). A smart parking system can be the first step in becoming a smart city. Besides solving many parking-related issues, it notifies users when parking time is up and when spaces become available. Additionally, the sensors can catch meter tampering, general malfunctions, and any issues with the electricity system.

3.4 Smart Grid

There are many large-scale civil IoT applications, but the smart grid has become the most significant one. With the help of IoT technologies, smart grids can supply robust and efficient energy management solutions that are lacking in current models/frameworks. Installing "smart" electricity meters with built-in sensors and IoT functionality can make it easier to check and control the flow of electricity, from industrial power stations to communal city blocks. In a smart grid, energy spikes and equipment failure can be detected, power outages can be prevented, and power can be routed more quickly to those in need.

In addition, the system provides the end-user with valuable information about their consumption patterns and how they can reduce or adjust their energy expenses. It is important to set up bi-directional communication between service providers and end-users to allow for the exchange of valuable information for the detection of faults, decision-making, and the repair of equipment (Singhal & Saxena, 2012).

3.5 Industrial IoT (IIoT)

Industry Internet, also called Industrial Internet of Things (IIoT), is the new buzz in the industrial sector and has appeared as the top IoT application area. What can IoT do for industries? Automation and machine learning, combined with IoT, help organizations reduce operating costs and increase productivity and efficiency. IoT can assist in modifying the packaging and design of the products to meet the needs of customer with high performance and satisfaction.

As a significant change, the internet of things offers solutions to all categories of its arsenal, including digitalization of industrial unit, monitoring of product development, inventory management, manufacturing, industrial processes, security, safety, quality assurance, quality analysis, optimization of packaging the products, supply chain management, etc. in the manufacturing sector, IoT can be used to check the efficiency of the systems, spot any errors in the machinery, and detect causes of inefficiency, etc. unplanned downtime can also be dealt with by IoT in the industry.

3.6 Connected Cars

With IoT technology developing rapidly, the automobile industry has undergone a radical transformation. Several dynamic changes have been brought about to the automobile due to its implementation, including improved safety, comfort, and luxury. The connected car gives a person the same connection, entertainment, and network that they have at their office, home, or at the entertainment center (Yang et al., 2018). IoT sensors embedded on the surface of vehicles further prevent accidents and ease a safe and smooth drive. AI is integrated into connected cars to provide alerts to drivers in dangerous situation in real-time such as lane departure, onward crashes and driving conditions. In-car entertainment, advanced navigation, maintenance features, telematics, fuel efficiency, and fleet management are just a few of the benefits that can be provided by this technology to the user.

3.7 Healthcare Sector

The use of IoT in healthcare can influence the sector in a positive way. IoT can be a valuable tool for physicians, patients, hospitals, and health insurance companies, all of whom are included in the healthcare industry.

The health of patients can already be enhanced and improved by wearables such as fitness bands and blood pressure monitors. Fitness bands and blood pressure monitors are just two of the wearable device's patients can use to improve their health. IoT devices provide doctors with continuous monitoring of metrics and automated alerts for vital signs, enabling them to supply better care to high-risk patients. The IoT device also makes it easy for physicians to look up the patient's history and access real-time health information. Even cancerous cells can be checked using nanotechnology based IoT solutions. Smart beds equipped with IoT sensors to recognize vital signs of the patients, monitoring the blood pressure, oximeters for body oxygen levels and increase and decrease in body temperature are another use of IoT technology. Hospitals can use IoT devices to track medical devices such as defibrillators and wheelchairs. They can also use them to manage inventory, check the environment, and regulate the temperature.

3.8 Smart Retail

Customers get a whole new shopping experience with IoT applications in retail. For retailers, IoT opens new ways to connect with customers and enhance in-store experiences. Using IoT applications, in-store checkout can be done quickly and efficiently. The checkout system using RFID (Radio Frequency Identification) reads tags on products and deducts the total amount from a customer's application, thus saving the customer time and frustration of having to wait in long lines.

Retailers can stay connected to their customers even when they are not in the store by using smartphones. Utilizing beacon technology and smartphones can help retailers better serve their customers.

3.9 Smart Supply Chain

If you use Amazon or Swiggy, you are probably aware of their tracking system. IoT applications are widely used in supply chains. By using a rating system, suppliers have been able to check goods in transit and receive immediate feedback from customers.

IoT systems can also supply information about the temperature and pressure at which items are being preserved, thus helping the supplier or driver keep the goods during transit. The system also allows clients to see the real-time status and a detailed view of the supply network.

3.10 Agriculture/Farming

We usually do not consider farming and innovative technology in the same light. Farmers, though, are already using these technologies to gain unprecedented access to data and the power to make decisions. IoT enables farmers to reduce waste and boost productivity. The IoT devices give farmers access to detailed information about their soil conditions, which is essential to growing healthy crops.

The characteristics of soil such as its moisture, acidity, nutrient concentration, temperature, and so on can aid farmers in planning irrigation, ensuring efficient water use, deciding the best time to plant, and even diagnosing soil and plant diseases. Data about short-term weather and climate can also be gathered and analysed using similar devices. Moreover, IoT devices can be beneficial to farmers by allowing them to track fleets, manage inventories, see fields, and even check livestock.

4. HARNESSING IOT TO TACKLE COVID-19

4.1 Practices Engaged in IoT Designed for COVID-19

The Society these days is facing a remarkable scenario. At the same time as each person feared a crisis in the traces of nuclear conflict, climate-related catastrophe, or comparable catastrophic threats, few imagined that a virulent disease should paralyze our international. It even gave birth to a brand-new ideato maximum people—'social-distancing.' suddenly, an awesome citizen is one that avoids public places and cares for loved ones whilst maintaining six toes in distance. As the wide variety of instances commenced growing and loss of life tolls increasing, technology like synthetic intelligence and the internet of factors have end up valuable equipment at some point of these challenging times. While the idea and the utility of artificial intelligence or ai are popular, IoT is an incredibly lesser-known idea. Covid-19 broke out and its particularly infectious nature turned into determined and healthcare specialists all around the world face the project of treating the diseased with minimum contact. This pandemic superior the modification and deployment of IoT gadgets to aid the healthcare sector (Rauch, 2021).

4.1.1 Track Quarantine

A crucial step to cut down the unfold of covid-19 is the powerful quarantine of infected or looked as if it would be infected humans. However, in a worldwide global, this is easier stated than executed. So, nations during the world grew to become to IoT and GPS enabled apps to track and, when essential, restrict such humans' movements. Russia, Poland, Singapore, south Korea are a few countries which can be going this path. Hong Kong commenced its quarantine efforts from the airport. Arriving passengers had been given wristbands at the side of a unique QR to tune their movements. Passengers downloaded an app known as 'live home safe' on their smartphones and scanned the qr. On attaining domestic, the man or woman had to walk around the condominium to calibrate the tool. The simple technology is geofencing, wherein a digital perimeter is created using GPS, RFID, wireless, Bluetooth sign, and mobile community.

4.1.2 Pre-Screening or Diagnosis

Even as hospitals and scientific facilities were quick to begin telemedicine services to diagnose and solution questions about covid-19, the variety of calls become overwhelming. Consistent with partner healthcare, Boston, the common wait time on their hotline peaked to half-hour, and lots of callers even dropped out inside this time. To counter this problem, software organizations collaborated with hospitals and clinical facilities to set up chatbots on their website and cellular apps. Those chatbots ask a series of questions to display traffic in keeping with the severity in their conditions. This way, the docs and clinical team of workers don't ought to answer the same questions time and again. They could as a substitute use this time to deal with sufferers.

4.1.3 Cleaning and Disinfecting

Cleaning, sanitizing, and disinfecting of clinical centers are indispensable and the infectious nature of covid-19 further emphasizes this step. Thanks to companies like TMiRob, UVD, and Xenex disinfection offerings, self-riding robots are used for this project. They disinfect the surfaces by using emitting excessive-intensity ultraviolet light, which destroys the virus by means of tearing aside their DNA. They may be wireless based and can be controlled through apps. Currently, those are being utilized in China, Italy, and America.

4.1.4 Innovative Uses of Drones

With social distancing turning into the brand new normal, drones have observed some revolutionary makes use of:

- To display and enforce the stay-at-home orders in Spain and China.
- To disinfect the exceedingly infected hotspot of Daegu, south Korea.
- To fly scientific samples and quarantine substances in Xinchang, China.
- To test temperatures of those in quarantine via infrared thermometers mounted on drones whilst the patients stand on their balcony.

4.1.5 Lowering Home-Grown Infections

there's an increasing consciousness among human beings to avoid touching susceptible surfaces like doorknobs, mild switches, and many others. Specially after touching mails or packages. Alternatively, they use IoT enabled smart speakers, lighting, protection systems, etc. To open doors and switch on lighting. This person currently used his clever security machine to request the transport man to depart the package deal internal his house while he unlocked the door from his cell phone. With the live-at-home orders in location, IoT offers us the power of video conferencing and surely assembly our loved ones with a simple voice command.

4.2 Worldwide Technical Innovations to Solve COVID-19 Issues Swiftly

As a result, to overcome and make the civilians more aware about the covid-19 pandemic, the government of India has released a telephone utility named as – Arogyasetu, that's aimed to expand a connection between the vital viable healthcare offerings and the people of India. Similarly, in China, the cell application called as – close touch is launched for its civilians. This application tells the app holder approximately the closeness to the corona-effective individual. So that the extra care can be taken even as moving outside. U. S. A. Government is soon going to launch a similar kind of cell utility for its civilians on the cease of April 2020. In early March 2020, a smart area medical institution became set-up in Wuhan, China, to supply a few remedies to the exhausted healthcare employees. This health facility was the proper example of a IoT, i.e., a mixture of IoT and ai. Robots and other IoT gadgets accomplished all the tasks on this medical institution—from checking temperatures of incoming sufferers to cleaning and disinfecting the place. They provided food and medicines to the quarantined sufferers and even entertained them through dancing. Every patient wore a smart bracelet and ring, so their vitals, consisting of temperature, heart rate, and blood-oxygen tiers, will be checked. If there have been signs and symptoms of infection, the clinical staff become intimated. Even docs and nurses wore those bracelets to capture early signs of infection. Several similar hospitals have been set up in Wuhan, that can accommodate nearly 20,000 patients if normal hospitals are crushed.

Presently, operations in Wuhan Wuchang smart field health center were placed on preserve due to fewer instances inside the location. According to a document by Forrester research, earlier than the covid-19 pandemic, most effective 7% of IoT solutions had been within the clever healthcare discipline in the Asia pacific region (APAC). Because the attain and severity of the pandemic improved, IoT started out taking a frontline in many nations, outside APAC, to control the disaster. Even though many have given you advert-hoc measures, it's time to offer concrete structures to most of these efforts, if now not all. Achim graze, an essential analyst at Forrester, gave a simple example of ways efforts can be taken a step similarly with IoT.

Many nations have installation temperature dimension systems at the doorway of maximum public locations like train stations and many others. If this accrued data (temperature) may be transferred and analysed in the cloud via an app, it is able to bring about real-time analysis. After China, Taiwan changed into the maximum predictable to have a greater number of instances of covid-19. However, Taiwan fast militarised and instituted unique methodologies for any viable coronavirus case identity, suppression, and resource provision to protect the fitness of the network. Taiwan furnished and included its countrywide health insurance database with its immigration department and took catalogue to instigate the introduction of big information for analytics; it generated real-time warnings at some stage in a clinical visit primarily based on journey antiquity and clinical signs and symptoms to useful resource case identity. They have additionally made use of this latest era, which incorporates scanning of QR code, related reporting of delivery records, etc. For the feasible identification of the inflamed ones.

4.3 Significant Applications of IoT for COVID-19 Pandemic

4.3.1 Wearable IoT Devices

Wearable gadgets which include bands, watches and even glasses were designed first for fitness and healthcare wishes, however they're swiftly turning into a great device in developing early diagnoses for covid-19. Wearables can without delay find whether an affected person is experiencing the onset of breathing problems that may be associated with the disease, then act quick to make a medical appointment earlier than more severe symptoms appear. In other cases, devices can ship a caution directly to a healthcare expert in order that steps may be taken proactively. Wearables are anticipated to turn out to be a $60 billion market by 2023.

IoT gadgets are supporting hospitals and clinics to diagnose patients from afar and prescribe far flung treatment. The faraway method is crucial during crises like the pandemic when hospitals want to best the time spent in character with sufferers that need immediate remedy. An excellent example is CMED health in Bangladesh, which has instituted an IoT-enabled fitness monitoring solution thru cell app. Crucial signs and symptoms are tracked in actual time, added first to a mobile device, then transmitted to the CMED cloud server. The app creates shade-coded diagnoses based totally on the extent of danger to the patient's fitness. Medical doctors had been able to locate and improve emergency situations during the covid-19 pandemic, and greater than 1. 5 million people have acquired assist from the system

4.3.2 Tracking Temperatures that may Possibly Signal Covid-19

Some other key IoT-enabled generation that is being used international to locate capability covid-19 contamination is infrared thermometers. We've all seen them in airports; however, they may be turning into pervasive anywhere to assist pick out an infected person inside a large crowd. Those who sign up a high body temperature are determined with the aid of clever sensors and can then be remoted or positioned into social distancing protocol. A current countrywide institute of health (NIH) study illustrated how the method works: IoT sensors consisting of infrared thermometers are placed in focused places, consisting of public lavatories, airports, buying shops, public transportation, hospitals, and workplaces, to name only a few. Sensors then wirelessly send information to regional gateway servers, which procedure and deliver to critical big facts and device studying infrastructure. Ai and deep learning techniques are applied to examine the healthcare trend, version risk institutions, and predict probably risky effects.

Scientists and authorities can then decide where an epidemic may be beginning and start protocol to mitigate the spread. IoT-related platforms are also being rolled out in places like amusement parks to discover potential covid-19 infections and check secure distancing approaches. Microsoft reviews that a related IoT framework known as the connected platform for detection and prevention uses a scalable smart facet and cloud infrastructure which can assist stumble on covid-19 in big organization areas. Detection components consist of:

a) thermal cameras and contactless thermometers to measure temperatures in humans
b) transportable virus testing facilities to conduct fast trying out for those with high temperature
c) smart cones to test social distancing in traces
d) automated announcements, indicators and reminders for employees and traffic introduced by means of interactive bots primarily based on IoT records
e) analytics to let authorities look at tendencies and effects

4.3.3 IoT Applications and Robots Keep Hospitals Safe

IoT-powered robots have grown in recognition in locations like hospitals throughout the covid-19 pandemic. Robots can be programmed to disinfect gadgets, smooth facilities, and even deliver remedy, reducing the need for human healthcare workers to behaviour those tasks and allowing them extra time to attention on treating sufferers. Furthermore, health facility directors can use IoT packages and video display units to greater without problems maintain tune of equipment including wheelchairs, defibrillators, and other essential generation that is vital in coping with and treating covid-19. IoT applications also are used in dealing with pharmacy inventory degrees and checking refrigeration and humidity degrees of medication to make sure protection.

4.3.4 Remote Patient Monitoring (RPM)

IoT-powered telemedicine enables hospitals and clinics to maintain faraway treatments. Rpm in particularises virtual technologies to accumulate health records from one patient and ship the data to healthcare vendors in a unique vicinity for assessment and guidelines. As an instance, CMED health (Bangladesh) gives an IoT-enabled fitness tracking solution through a cell app. Customers can test their number one health vitals remotely by means of integrating IoT-enabled clever clinical gadgets with the app. The measured data is sent to CMED's cloud server, which may be accessed and analysed by way of authorized medical doctors. The CMED app produces colour coded results based at the emergency degree' of the consumer's health reputation. During covid-19, this allowed CMED's crew to find and increase emergency situations. 1. 5 million people in Bangladesh have received assist from CMED fitness's platform at some point of the pandemic.

4.3.5 Healthcare Management

Poor healthcare control in developing nations is regularly connected to the lack of efficient visualisation of medical institution ability, mainly with regards to mattress availability. To reply to this shortage, south Africa-based Gauteng health services added a digital bed management gadget (eBMS) to discover the provision of beds throughout multiple websites. The use of cloud-based totally generation, the IoT sensors placed at the beds allow health facility team of workers to seamlessly find the beds' availability. eBMS utilization led to vital reductions within the wait time for a bed, offering patients in emergency departments with spark off get entry to the care. IoT answers together with eBMS can give essential steerage to healthcare stakeholders and assist governments put together for future pandemics.

4.3.6 Vaccine Cold Chain Monitoring

Making sure essential immunisation services at some point of covid-19 has verified to be tough in developing nations. Mobile era and IoT can optimise the vaccine's deliver chain. Through IoT sensors placed at the vaccine, bloodless chain facts loggers send correct facts of condition logs through cellular statistics networks to the cloud. One example is the digital vaccine intelligence network(eVin), an IoT-enabled cell-based technology developed via the UNDP and the Indian government, elements actual-time coordination control across the vaccine bloodless chain. The app – related to IoT sensors positioned on the vaccine – tracks the place, temperature and stock tiers of vaccines, ensuring the deliver is safe and reliable. The adoption of eVin in India has resulted inside the reduction of vaccine stock-outs through eighty per cent.

4.3.7 Healthcare Delivery Drones

IoT-enabled drones have proven to be a lifeline for the delivery of tests, PPE, drug treatments and different crucial clinical elements to populations in growing countries. Since can also 2020, zipline enabled drones to deliver essential clinical materials to rural fitness centres in Rwanda and Ghana. The drone company can provide approximately one hundred sixty distinct clinical products, serving near two thousand five hundred hospitals and health facilities throughout Rwanda and Ghana during the pandemic (Chebib, 2020). Other styles of drones played a lively position in disinfecting public spaces or detecting covid-linked symptoms.

5 MERITS OF USING IOT TECHNOLOGY FOR COGNITIVE ASSISTANCE

5.1 Management and Automation

Although some may describe it as not anything extra than a gimmick, the common-or-garden, smart bulb is one of the great and most effective examples of manipulate and automation via an IoT tool. Smart light bulbs have exploded in popularity in latest years for a reason (Vishwakarma et al., 2019). At the face of it, those gadgets, which connect to your home community, offer you with nothing greater than the capacity to show the lighting in your property on and off remotely. But a clever bulb is extra than only a far-off mild switch. It enables you to set the scene for a particular time of day, routinely flip the lighting fixtures off within the dwelling room whilst you open Netflix on your smart television, and so forth.

Manifestly, control and automation are internet of things benefits that span beyond clever mild bulbs. Internet-linked gadgets including air-conditioners, coffee makers, and humidifiers can all be integrated with google- and amazon-made voice assistants, permitting you no longer simplest to turn them on or off via easy voice commands but additionally to automate their schedules. In relation to enterprise and industrial applications, IoT devices can be used to control robotic assembly systems, production traces, printers, and all forms of different machines in an office area or manufacturing unit.

5.2 Real-Time Access to Data

One of the number one net of factors advantages is the uninterrupted go with the flow of data. Internet-linked gadgets can share statistics at the velocity of mild, this means that fewer delays and a decrease capability for miscommunication. Consider a huge warehouse wherein every product is logged and related thru an IoT community. Information such as product movement could be updated in real-time, permitting new inventory to be ordered robotically at the precise proper time and in the specific correct quantity. Real-time monitoring is even greater crucial in different settings, which include nuclear energy flowers or maybe grain silos. Sensors measuring temperature or humidity may be used to inform an IoT gadget, which could, in flip, routinely manage these parameters.

5.3 Promising Business Outcomes

The rise of present-day era constantly brings an explosion of new enterprise opportunities, and IoT is no distinctive. Although clever gadgets have lengthy penetrated almost every issue of society, we are still years, or possibly decades faraway from huge international use. Internet of things software improvement offerings supplied through Sumatosoft are at the leading edge of technological development, providing companies with next-era business IoT solutions. Many other corporations are revolving around the usage of IoT, consisting of putting in internet-related devices into cars, building smart wearables designed explicitly for humans with excessive blood strain or diabetes, and so forth. The internet of factors may also doubtlessly be used by branding and advertising and marketing organizations, which could use artificial intelligence and advanced data series to expand their virtual method offerings. This feeds perfectly into the subsequent object on our internet of factors blessings list.

5.4 Data Collection

The IoT resources remarkable possibilities for statistics series and commercial enterprise growth. Deployed across a wide range of devices, the internet of things advantages of statistics collection can assist business proprietors make choices on a macro scale as well as a granular stage. An easy example would be net-linked tags on retail products, that can supply actual-time data approximately buying choices and developments on a daily or weekly foundation. This relatively trustworthy system could enhance stock manage and offer plenty-wished perception into patron behaviour and purchasing patterns (Zubovich, 2022). Move-referencing those figures with other information, such as zip codes or membership card numbers of person clients, can produce large swimming pools of statistically applicable statistics.

5.5 Enhanced Efficacy

Via its very definition, the internet of things runs without human intervention. It's far a gadget relying totally on device-to-gadget communique, with statistics logged and accumulated in real-time, at any time of day or night-time. One of the most massive effects of such a system is the multiplied efficiency of a huge variety of services. Staff contributors no longer need to spend treasured time gathering and processing records, as machines can efficaciously take over this job. This internet of things benefits leaves human workforce greater time for innovative endeavours, permitting them to awareness on using the accrued data rather than the collection of data itself.

Monitoring routes and mileage of cars is an ordinary example of machine-to-device conversation improving efficiency. A massive company that owns a fleet of automobiles can enchantment to employer software improvement businesses and construct a device with the intention to song the mileage and utilization of each automobile using IoT gadgets in preference to rely on employees to complete this task. The devices would then send facts back to a server in actual time to song and test automobile use.

5.6 Enhanced Quality of Life

To date, we've dealt mainly with the internet of things benefits in commercial enterprise however focusing completely on this factor could be quick-sighted. There are many approaches wherein massive statistics and IoT ought to improve the first-rate of existence of ordinary humans. As an example, in remedy, clever devices may want to store many lives. Clever tablets sending lower back records as they make their way along the gastrointestinal tract or blood-stress video display units sharing data with an IoT system in actual-time may be authentic lifesavers. Advanced pleasant of lifestyles from IoT networks doesn't handiest follow to individuals. Smart gadgets could help entire groups, with internet-related site visitors' sensors and lighting easing congestion in closely populated regions, as an example. Clever gadgets in automobiles should connect not most effective to traffic lighting however additionally to toll gates and street safety monitors, offering drivers with actual-time feedback on avenue situations at the way to their respective destinations. Of direction, smart metropolis traffic requires highly priced logistics & transportation software development services, but it will repay over the years.

5.7 Cost Reduction

interconnected gadgets deliver many opportunities for fee discount. The IoT mesh no longer simplest resources a chance to analyse statistics in real-time but improves efficiency across the board, as explained in advance. This will offer a spread of internet of factors advantages, along with extra effective inventory management, discounts in development price, or even reduced costs in manufacturing and shipping of uncooked materials. As businesses turn out to be smarter, fees must theoretically move down across the board. Ambient temperature sensors could be linked to computerized cooling and heating structures, providing exceedingly efficient weather manipulate without a human interplay by any means. Lights may be installation to turn on simplest, when essential, for that reason lowering utility costs, and so forth. Cost discount is one of the strongest effects of internet of factors on businesses and one which motivates commercial enterprise proprietors to put money into it.

5.8 Higher Productivity

IoT also can enhance performance and productivity at paintings and within the commercial sector, like how voice-activated amazon and google merchandise increase performance at domestic. Facts gathered thru a community of IoT gadgets, as an example, elements distinctly precise statistics that may be used by corporations to improve their inner methods and customer service. As a sensible example of the internet of factors blessings, if an energy company cannot advantage get entry to a belonging, it will estimate its software bill. However, if the electric box has been linked to the utility organization's server thru the net, the customer could by no means should worry approximately an overcharged anticipated invoice once more.

5.9 Big Data and Predictive Analysis

Right, here's a buzzword that you've nearly honestly heard but won't recognise lots approximately — huge facts. This word has been around long earlier than IoT took off, and it describes a process of gathering and analysing considerable quantities of facts for numerous purposes. One of the purposes of IoT systems is to acquire records, whether to song customer interest, inventory, automobile use, temperature, or whatever else you can consider. All this information is fed returned into the gadget, in the end improving the efficiency of diverse techniques. Successful evaluation of massive records can predict the whole lot from inventory motion to purchaser behaviour and generate new possibilities inside the commercial enterprise international. As increasingly more smart devices reach our houses and colleges and more facts is collected, we'll see an ever-growing variety of net of factors advantages and methods wherein predictive strategies can be used to enhance our lives.

5.10 Health and safety

Technological and clinical advances have enabled more people to attain old age, however those obstacles are being driven in addition each yr. In a medical emergency, each 2nd may want to mean the distinction among life and demise, and the quite interconnected world of IoT ought to result in a plethora of advantages. Believe a simple fitness bracelet with a system to preserve a watch on blood sugar tiers. This net-linked tool ought to log any blood sugar or insulin stages changes and ship notifications directly to scientific employees or circle of relative's individuals. For human beings with chronic illnesses, who should preferably be constantly checked via fitness specialists, discreet IoT gadgets may want to supply lifestyles-saving real-time records. 6 IoT for healthcare: challenges and opportunities (Rasool & Dulhare, 2022).

6 IoT FOR HEALTHCARE: CHALLENGES AND OPPORTUNITIES

6.1 Several IoT Proposals are Unfinished or Fruitless

comfort and rapid statistics transfers are two issues that could inspire health care groups to discover IoT technologies. There's absolutely purposed to be excited about IoT's capability. But 2017 studies from cisco, painted a less-than-glamorous picture of IoT transformation efforts. The studies concerned getting comments from extra than 1,800 humans across the USA, United Kingdom, and India, who have been stakeholders in past or ongoing IoT tasks. Cisco's survey found out that finished projects were only considered successful 26 percent of the time.

Moreover, about one-1/3 of respondents taken into consideration their completed tasks unsuccessful. Most tasks—60 percentage—stumble upon hassle at the evidence-of-idea stage or rapidly thereafter. But it's worth noting that utilising external partnerships (e. G., platform companions) became an essential element for the ones businesses that finished a hit implementation. Even though this cisco looks at didn't cognizance its evaluation on health care agencies, that industry vertical was represented in the typical survey. The findings emphasize that agencies need to be careful whilst making plans their IoT rollouts in 2019. For example, they must start small and prioritize initiatives that align with their most distinguished enterprise aims or affected person wishes (Matthews, 2018).

6.2 Plethora of Data Generation from Healthcare

Several the most interesting fitness care IoT tasks encompass ways to reduce emergency room ready times, music property and those that move for the duration of hospitals and offer proactive indicators approximately medical gadgets which can quickly fail. The ones improvements are certainly incredible, however one task related to all of them is the amount of data generated. A forecast indicates that by 2025, health care will cope with producing the most records of another zone.

Now is the time for organizations to understand that identifying to apply IoT technology will likely make data storage wishes would pass up. They'll observe the differences as soon as 2019. Moreover, the fitness enterprise must be tremendously cautious to deal with patient statistics from IoT devices in keeping with federal and state policies. The flood of information created via the IoT devices and gadgets used in the health care enterprise may also reason unforeseen problems if agencies are not prepared to address it well and affirm its first-class.

6.3 IoT Devices Increase Available Attack Surfaces

the vast possibilities for the use of IoT devices in fitness care also gift regarding vulnerabilities. As tool use rises, so does the variety of ways hackers could infiltrate the machine and mine for the most precious facts. One new threat a Zingbox examine found is that hackers could study how a linked scientific tool runs by means of entering the machine and studying its errors logs. The understanding the hacker's advantage ought to ease breaking right into a hospital community or making devices submit incorrect readings that influence patient care.

On a more advantageous be aware, but Zingbox's research showed progress in carriers, vendors, and producers' willingness to collaborate. Those shared efforts may want to reduce patient dangers by using ultimate the gaps that can form among the layers of an IoT system by using reinforcing standards and normalizing secure protocols. It's not feasible to recognize all the cybersecurity dangers fitness agencies may additionally face in 2019. However, centers planning to put into effect IoT generation should take care to increase awareness of current threats and recognize the way to defend networks and devices from hackers' efforts.

6.4 Outdated Infrastructure Hinders the Medical Industry

Even though retrofitting can breathe new existence into growing old infrastructure, surely taking benefit of IoT is tricky if a facility's infrastructure is outdated. Vintage infrastructure is a known trouble in health care. Whilst hospitals are in dire want of made over infrastructure, additionally they have trouble hiring the team of workers to make improvements. Tech skills is in excessive demand. Prospective candidates may not need to address vintage infrastructure.

6.5 IoT Poses Many Overlooked Obstacles

In accordance to investigate from Aruba Networks, the maximum not unusual use of IoT generation in fitness care is to use it to affect person tracking structures. It's undoubtedly handy to take that technique, but something fitness groups frequently overlook is that not like websites, as an instance, those gadgets generally can't go through planned durations of downtime. As a substitute, updates ought to arise continuously as people use the monitoring gadgets. Moreover, hospitals often depend upon IoT-enabled deliver cabinets to tune assets. Once those systems are in region, the centers can frequently lessen earlier stock control troubles, however even the best linked gadgets can't remove human mistakes. In the end, we make those IoT structures. If people are blunders-prone, it follows that our IoT systems can inherit blunders-prone behaviour from us (Dulhare & Rasool, 2019).

Supplier tests also are essential to conquering often-unanticipated demanding situations. Some producers are typically concerned with beating competitors to the marketplace with their products. Inside the rush, many don't construct protection into their tactics from the start, so they shouldn't be amazed whilst patron databases are breached. Even though a health center has above-average cybersecurity defences, sufferers may also still be at risk from products that lack adequate safety. One horrific apple can smash the bunch. Cybersecurity must be uncompromising and complete. Regrettably, few modern IoT systems are secure by conventional network protection metrics.

7. HEALTHCARE TO IOT SECURITY BEST PRACTICES

Protection breaches may want to take place thru lack of knowledge, negligence, or sick cause. In case your employer falls victim to a hacker attack through a vulnerability in a connected tool, numerous events will be held accountable: your staff, cloud provider providers, patients, or regulatory our bodies not paying closer attention to a capability trouble. A preventive method but works high-quality on an organizational level. Underneath is an outline of the exceptional practices aimed toward cutting the cybersecurity demanding situations in using IoT in healthcare.

7.1 Ensuring Network Security

Healthcare groups keep away from ability breaches in network protection by offering network segmentation and protective every of the subnets at its personal degree. Community directors can execute manage over the go with the flow of traffic between every community section and use encryption strategies to shield records from being decoded, although it gets intercepted by way of hackers.

7.2 Applying Context-Aware Security Approach

Context-conscious security structures are, in essence, greater advanced than traditional cybersecurity tools detecting and preventing the already-acknowledged threats. They can belong to ai-driven protection systems. Contextual-aware safety considers the broader photograph – who is trying get entry to, from in which, while, and in what way (Kumar et al., 2018). By way of detecting non-traditional hobby and surprising styles, such a method facilitates locate threats in actual-time and save you security violations and community healthcare IoT breaches.

7.3 Device Centralization and Segmentation

healthcare IoT protection excellent practices consist of aggregating them into a separate network to ease their tracking and manage. IoT aggregation hubs will help you manage your gadgets, control which community elements and assets they've get admission to and exchange their protection settings.

7.4 Protection on a Hardware Level

As stated above, applying data encryption is also part of the healthcare IoT safety best practices. Mostly, connected devices use each symmetric and asymmetric lightweight cryptography (LWCRYPT) strategies, while events alternate encryption keys before conducting data switch. Today, IoT sensors typically have encryption keys geared toward setting up a covered http channel between gadgets and consumers.

7.5 Data Encryption

As mentioned above, applying data encryption is also part of the healthcare IoT security best practices. As a rule, connected devices use both symmetric and asymmetric lightweight cryptography (LWCRYPT) techniques, while parties exchange encryption keys before conducting data transfer. Today, IoT sensors normally have encryption keys aimed at setting up a protected HTTP channel between devices and consumers (Saraswat et al., 2019).

7.6 EMI Shielding

EMI stands for 'electromagnetic interference trouble', that is turning into a part of our lives with increasingly more electronic gadgets. Shielding involves constructing a metal frame surrounding a device and blocking off electromagnetic waves to defend it from unwanted interference.

7.7 Visibility Maintenance

Connecting new devices means extra security challenges of IoT in healthcare for their tracking. The equipment number affects the complexity of IoT healthcare security. The hardships of monitoring require using healthcare visibility solutions to track and trace mechanisms. That is why you need to configure a relevant IoT visibility application to do that efficiently at the very start of the devices' work.

7.8 Protection Against Malware and Trojan Threats

Several protection methods are at hand for efficient IoT security. The most popular of them are:

- Signature-based detection works with the antivirus system's signature. Malware can be detected only when the database signatures don't coincide with the scanned one. Only IoT devices with a small memory can use this method effectively.
- Static methods use devices' static characteristics. The static analysis uses different tools for the simple signature collection and identification. This practice allows malware searches without the actual code change. The static approach is limited in verification but cost-effective and easy to use.
- Dynamic methods of detection see suspicious activities. They track changes in network behaviour, CPU load, virtual memory, calls, and SMS.

The best way to protect your software from trojan, malware, and hacker attacks is to use a combination of mentioned above methods (Acharya et al., 2021).

7.9 Use Proven Norms and Best Procedures

There are specific approaches that enable reliable medical healthcare IoT security systems. We can highlight obligatory encryption, firewalls, and hard-coded passwords elimination among such techniques. Regular devices updates are also mandatory. Moreover, medical providers need to assess each device to be conscious of healthcare IoT vulnerabilities and detect any suspect network traffic. They can use behavioural analytics profiling, for that matter.

7.10 Utilize Right Instruments

Security optimization requires using specific applications to streamline the reliable work of IoT healthcare devices. Various platforms can automate control of substantial data amounts, devices' management, and handle authentication certificates. Specific medical equipment control tools created by different manufacturers give information about the device: find it, what data is collected from it, and where its internet connection is set up. Administrators use those tools to check internet traffic and manage network connections to approve or deny them.

REFERENCES

Acharya, J., Chuadhary, A., Chhabria, A., & Jangale, S. (2021). Detecting Malware, Malicious URLs and Virus Using Machine Learning and Signature Matching. *2021 2nd International Conference for Emerging Technology (INCET),* 1-5. 10.1109/INCET51464.2021.9456440

Baidya, S., & Levorato, M. (2018, September). Content-Aware Cognitive Interference Control for Urban IoT Systems. *IEEE Transactions on Cognitive Communications and Networking, 4*(3), 500–512. doi:10.1109/TCCN.2018.2815604

Chebib, K. (2020). *IoT applications in the fight against COVID-19.* https://www.gsma.com/mobilefordevelopment/blog/iot-applications-in-the-fight-against-covid-19

Dulhare & Rasool. (2019). IoT Evolution and Security Challenges in Cyber Space: IoT Security. In *Countering Cyber Attacks and Preserving the Integrity and Availability of Critical Systems.* IGI Global.

Dulhare, U. N., & Rasool, S. (2022). Smart Airport System to Counter COVID-19 and Future Sustainability. In C. Satyanarayana, X. Z. Gao, C. Y. Ting, & N. B. Muppalaneni (Eds.), *Machine Learning and Internet of Things for Societal Issues. Advanced Technologies and Societal Change.* Springer. doi:10.1007/978-981-16-5090-1_5

Fossum Færevaag, E. (2021). *IoT's Evolution During COVID-19 and Into the 'New Normal'.* Channelfutures. https://www.channelfutures.com/iot/iots-evolution-during-covid-19-and-into-the-new-normal

Founoun & Hayar. (2018). Evaluation of the concept of the smart city through local regulation and the importance of local initiative. *2018 IEEE International Smart Cities Conference (ISC2),* 1-6. 10.1109/ISC2.2018.8656933

Hurford, R., Martin, A., & Larsen, P. (2006). Designing Wearables. *2006 10th IEEE International Symposium on Wearable Computers,* 133-134. 10.1109/ISWC.2006.286362

Kim, J. Y., Lee, H., Son, J., & Park, J. (2015). Smart home web of objects-based IoT management model and methods for home data mining. *2015 17th Asia-Pacific Network Operations and Management Symposium (APNOMS),* 327-331. 10.1109/APNOMS.2015.7275448

Kumar, S., Shanker, R., & Verma, S. (2018). Context Aware Dynamic Permission Model: A Retrospect of Privacy and Security in Android System. *2018 International Conference on Intelligent Circuits and Systems (ICICS)*, 324-329. 10.1109/ICICS.2018.00073

Matthews, K. (2018). *5 Challenges Facing Health Care IoT in 2019*. Iotforall. https://www.iotforall.com/5-challenges-facing-iot-healthcare-2019

Mazumder, D. (2021). A novel approach to IoT based health status monitoring of COVID-19 patient. *2021 International Conference on Science & Contemporary Technologies (ICSCT)*, 1-4. 10.1109/ICSCT53883.2021.9642608

Rasool & Dulhare. (2022). Data Center Security. In *Green Computing in Network Security: Energy Efficient Solutions for Business and Home*. CRC Press. doi:10.1201/9781003097198

Rauch, S. (2021). *Using IoT Applications to Manage the Spread of Covid-19*. Simplilearn. https://www.simplilearn.com/using-iot-applications-to-manage-the-spread-of-covid-19-article

Saraswat, P., Garg, K., Tripathi, R., & Agarwal, A. (2019). Encryption Algorithm Based on Neural Network. *2019 4th International Conference on Internet of Things: Smart Innovation and Usages (IoT-SIU)*, 1-5. 10.1109/IoT-SIU.2019.8777637

Simon. (2021). *The Internet of things (IoT). The Internet of things (IoT) describes....* https://simonbrard017.medium.com/the-internet-of-things-iot-describes-the-network-of-physical-objects-so-known-as-things-cb8c9c994603

Singhal, A., & Saxena, R. P. (2012). Software models for Smart Grid. *2012 First International Workshop on Software Engineering Challenges for the Smart Grid (SE-SmartGrids)*, 42-45. 10.1109/SE4SG.2012.6225717

Vishwakarma, S. K., Upadhyaya, P., Kumari, B., & Mishra, A. K. (2019). Smart Energy Efficient Home Automation System Using IoT. *2019 4th International Conference on Internet of Things: Smart Innovation and Usages (IoT-SIU)*, 1-4. 10.1109/IoT-SIU.2019.8777607

Wilford, E. (2022). *10 IoT technology trends to watch in 2022*. Iot-Analytics. https://iot-analytics.com/iot-technology-trends

Yang, T., Wolff, F., & Papachristou, C. (2018). Connected Car Networking. *NAECON 2018 - IEEE National Aerospace and Electronics Conference*, 60-64. 10.1109/NAECON.2018.8556715

Zubovich, N. (2022). *Advantages of Internet of Things: 10 Benefits You Should Know*. https://sumatosoft.com/blog/advantages-of-internet-of-things-10-benefits-you-should-know

Section 2
Affective Computing

Chapter 7
Feasibility and Necessity of Affective Computing in Emotion Sensing of Drivers for Improved Road Safety

R. Nareshkumar
Department of Networking and Communication, SRM Institute of Science and Technology, India

K. Nimala
Department of Networking and Communication, SRM Institute of Science and Technology, India

G. Suseela
iD https://orcid.org/0000-0002-9162-7444
Department of Networking and Communication, SRM Institute of Science and Technology, India

G. Niranjana
Department of Computing Technologies, SRM Institute of Science and Technology, India

ABSTRACT

The development of the automobile industry and civilian infrastructures improved the lifestyles of everyone in the world. In parallel to the rise in quality of life of everyone, the number of road accidents also rose. The major reason behind road accidents is emotional factors of the drivers. The emotional imbalance will influence the drivers to abandon the traffic rules, neglect speed limits, cross the signals, cross the lane, etc. Recently automobile industries have extended their researches to the development of emotion sensing systems and embedding them inside the vehicles using affective computing technology to mitigate the road accidents. These emotion sensing systems will be decisive and act as human-like driver-assistive systems in alarming the drivers. This chapter focuses on bringing out the feasibility and existing challenges of affective computing in sensing the emotional factors of drivers for improved road safety.

DOI: 10.4018/978-1-6684-3843-5.ch007

Copyright © 2023, IGI Global. Copying or distributing in print or electronic forms without written permission of IGI Global is prohibited.

1 INTRODUCTION

We are currently through an information revolution during which technology improves people's work and daily activities, allowing them to be more productive; however, it is critical to assess whether the progress of technology is actually beneficial to people. In some circumstances, technology is not always effectively embraced; for example, if something new or novel is misappropriated, resulting in disinformation and misunderstanding, it could have the opposite effect. (Morley & Parker, 2013).

Emotions are important in almost every aspect of our everyday lives, including decision-making, motivation, and interpersonal interactions (Eyben et al., 2010), and driving is no exception (Jeon, 2016; Jeon et al., 2011). Some of the most common emotional triggers include a loss of control, journey delays, the risk of accidents, and the greater intellectual load required. Drivers who rely substantially on riding as a based-on job activity may be more susceptible to these triggers. (e.g., cab drivers, package transport). Although little stress might assist people in achieving their objectives, such as being on time at their destination, too many or too few factors can have a significant impact on driving effectiveness and general well-being (Ding et al., 2014).

As a result, future vehicles that can detect and respond to the emotional states of riders and drivers would be successful in improving not only highway safety but also mental health. Recent technological advancements, such as portable tech, have made it possible to investigate emotions in the real world, resulting in a slew of publications studying the harmful effects of particular emotions while driving.

1.1 Sensing and Pre-Processing

1.1.1 Face and Head

Facial and head actions are being used to infer driver sentiment, primarily by tentative facial expressions (e.g., laughing, scowling) and head signs (e.g., signals, angles) while driving. Traditional RGB cameras have been used in most of the experiments to capture expression and head information (e.g., (Ma et al., 2017; Paschero et al., 2012)). There are a few less commonly investigated techniques, such as thermal and ir cameras (Guo et al., 2014; Kolli et al., 2011), which may be more resistive to certain sorts of brightness variations. The world's most important way to accurately detect the image quality of facial emotions is to look at the driver from the front. In controlled laboratory investigations, researchers usually place the camera on high of a monitor or simulator to gain a frontal view (Agrawal et al., 2013; Moriyama, 2012). The sensor has been put on the vehicle top in far less regulated conditions, however it may partially obscure the driver's view (Cruz & Rinaldi, 2017). Another option is to use the car display, albeit the camera vision may be blocked by the gear stick during turns (Guo et al., 2014).

Facial and head motions are used to infer driver emotion, usually by observing facial features (e.g., happy, angry) and head movements (e.g., shrugs, tilts) while driving. Traditional RGB cameras were generally used in the studies to obtain face and head data. IR cameras and surveillance equipment (Guo et al., 2014; Kolli et al., 2011) are less commonly used techniques that may be more resistant to certain sorts of illumination fluctuations. A frontal perspective of the driver is usually favored to accurately identify the range of facial expressions.

To get a frontal image in controlled laboratory trials, researchers commonly set the camera on top of a monitor or simulator(Eyben et al., 2010). Researchers have mounted the camera on the automobile windshield in less controlled circumstances, while it may partially obscure the driver's view(Cruz & Rinaldi, 2017). The automobile dashboard is another option, however the camera image may be partially obstructed by the rear wheel during turns(Guo et al., 2014).

1.1.2 Speech

In a chaotic and fluctuating engineering situation, the signals produced by the vocal chords that may be related to the mental situation of the driver, such as the speed and volume of a voice.

The research looked at directional microphones (Tawari & Trivedi, 2010), condenser speakers (Schuller et al., 2006), and sensor arrays, among other things(Jones & Jonsson, 2008). The arrangement of the microphones is critical for obtaining important data while reducing vehicle and environmental noise Different filters are routinely used to audio streams after they have been collected to motivate and inspire relevant sound and reduce other overlapping noise. (e.g., machine of the coach).

1.1.3 Behaviour

Few lessons used malicious judgments across several documents sectors to examine the association between distinct behavioral changes and emotions (e.g., (Oehl et al., 2011; Karwowski et al., 2010)). For example, (Karwowski et al., 2010) (Oehl et al., 2011) found that usual hold strong point differed ominously when people were angry or happy. Furthermore, stress may be measured using the directing position and a frame coil check model. Furthermore, other educations looked at the relationship between various emotion annotations and behavioral driving characteristics.

1.1.4 Affective Facial Identification' Barriers

Facemask evaluation is considered one of the furthermost basic capabilities the human brain, has up till now to be fully integrated in a machine. Challenges to achieve the objective in the past have centered on enhancing the efficiency of emotion classification tests. The importance of revealing the distinctiveness of affective categorization was overlooked. Depending on the context, geographic, and temporal signals, the same facial expression can be perceived in a variety of ways. In a summary, this leaves affective classification uncertain.

2 DRIVER EMOTION RECOGNITION

Methods for recognizing emotions rely on data from one or more sources. Methods that analyses facial expressions, for example, procedure video camera metaphors, whereas approaches that analyses biological reaction of the social body use data from sensing devices(Wrobel, 2018). Single-modal emotion detection relies on a single emotional component to recognize emotions, which has significant limits(Bai et al., 2018). As a result, several academics have looked into bimodal emotion detection using single-modal recognition as a starting point.

2.1 Human Action Recognition (HAR)

Human Action Recognition (HAR) is the attempt to analyses human activities using complex procedures such that a machine may perceive, evaluate, explain, and classify them given any acceptable input or input (Sarkar et al., 2021). The human motion video, which includes hand waving, strolling, racing, clapping, and boxing, is first converted into a 2D picture, which is then preprocessed, proceeded by image retrieval using LST, and classification using the KNN classifier.

Humans are good at recognizing and identifying activities in video, but automating this process is difficult. Human action detection in video is of interest in applications like as automated surveillance, senior efficiently in order, human-computer interaction, information video lookup, and video summarization.

2.2 Categories of HAR

Based on Sensors Sensors are either directly or indirectly linked to the human subject's body or the surrounding area to collect essential input of motion patterns and input signal reaction data (Sarkar et al., 2021).

Visual sensors, like as RGB sensors, are commonly employed to capture visual information in form of two-dimensional images (Sarkar et al., 2021).

2.3 Expression Label Characteristics

This subsection describes the statistics of the Affective algorithm's surface sentiment tags, as well as the de - duplication, pre-processing, and transformation techniques that were applied to them (Zepf et al., 2021).

2.3.1 Emotion Annotation Tool

Its emotion detection system estimates emotions image - based using pattern recognition and deep learning models. Unlike other facial expression recognition solutions, Affective's procedures remain based on a massive introductory dataset with over 9 million facemask pictures from 85 countries, over 4 billion visual structures, and seven years of audiovisual footage.

2.3.2. Emotion Persistence

Facial terminologies be situated frequently short-lived, continuing among 0.5 to 4 seconds. Affective Facial Analysis's Difficulties To guarantee that the driver's highly frequent and stable affective states were adequately captured, we had to use a non-overlapping sliding window and segmented the driving data into driving segments. The emotion labels were assigned to each driving section based on their average level of present. Our goal was to estimate driver emotion using CAN-Bus data . information from front-view photography of the identical riding segments.

2.3.3 Emotion Distribution

It's crucial to look at the brand supply to see in what way accurate we container forecast sentiments (N. K. & J. R., 2019; Nimala et al., 2020)and to make sure our powdered truth is correct ().

It can be used to make the following conclusions:

Extra undesirable than confident valence was elicited by the drivers
across diverse drivers, a unique customized mean value for each emotion
The majority of emotions have a modest level of presence

Figure 1. Facial Emotion Recognition

Figure 2. Effects on the human body and a sentiment classification model during driving.

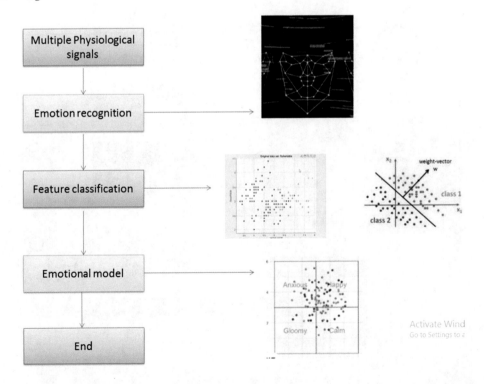

3 HUMANMACHINE INTERFACE (HMI)

Driving is becoming more complicated, and the purpose of information management is evolving at a rapid rate. Previous in-vehicle communication systems were designed to warn drivers of potentially dangerous situations without overloading them with information (Davoli et al., 2020).For this case inattention and workload are the most important concerns to consider when evaluating an HMI from a human standpoint. The modalities for representing information and the timing for displaying it are hot subjects in modern HMI design. HMI design is increasingly resolving this transition by focusing on interaction rather than presentation.

In more detail, the encounter anticipates more active user participation, necessitating available options. User-Centered Design (UCD) has become the main method for adapting HMI development around users in recent years, for example, in the automotive sector.

Emotion control solutions have lately been considered in order to increase interaction performance.

Figure 3. Human Emotion Recognition Model.

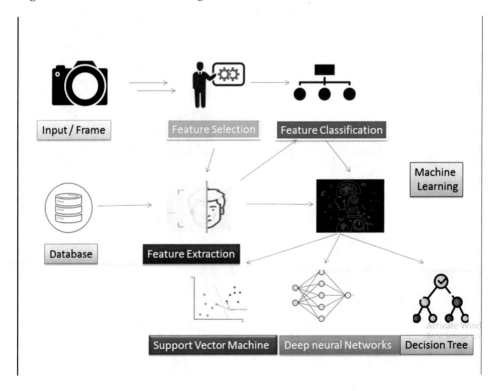

4 HUMAN EMOTION RECOGNITION
WITH SENSING MATERIALS

When it comes to reviewing and evaluating breakthroughs for driver computer vision and human monitoring in general. It's vital to note that the final user's approval of these technologies is mostly decided by their unobtrusiveness or the device's lack of apparent privacy intrusion Multicultural sensors, such as fixed sensors and mobile sensors, can be employed as observation ways to attain this purpose.

However,(Davoli et al., 2020) trained models may miss detect items or fail to detect them entirely. The accuracy of the three components is crucial to our recognition process (The eye pair, the nose, and the mouth). When it comes to monitoring the facial components, miss occurrences cannot be accepted. The components of the component-based face recognition system must be cropped and displayed appropriately. Miss detection may cause the learning process to represent irrelevant data (as seen in Figure 1), resulting in a decreased recognition success rate. The most important of the out the components is the eye pair. The eye pair contains the majority of a person's facial information. It's also the benchmark object for detecting and cropping the rest of the facial element in this technique.

Figure 4. Sensing Component of Human Emotion Recognition Model.

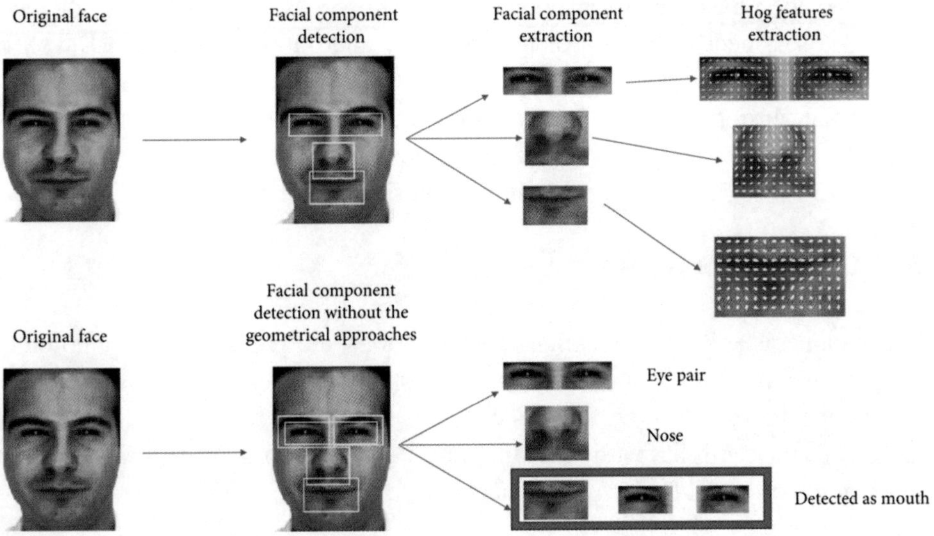

101

5 METHODOLOGIES

We initially created selection criteria for locating works fixing the issue of feeling assessment and/or study in the environment of driving in order to reduce familiarity bias and locate relevant publications to include(Zepf et al., 2021). We omitted studies that focused solely on drivers' mental and physical conditions, such as drowsiness/ fatigue, attention problems, and mental workload, because these topics have been widely covered by previous research.(Davoli et al., 2020)

The first step was to create relevant driving scenarios. To strengthen the applicability of the results, agricultural, urban, and highway driving scenarios were developed, as these are the most prevalent settings faced by a driver. The another phase entailed carefully selecting driver participants in order to ensure that the test group accurately reflects the overall population. The third stage involved computation in order to produce useful results. The fourth stage involves identifying differences in vehicle safety and comparing them to observable traits between ADAS-equipped and non-ADAS-equipped vehicles.

5.1. Inertial Sensors

Inertial Measurement Units (IMUs) are used to estimate the mobility level of mobile sensing devices, which are commonly worn by the person being watched. Because IMUs include 3-axis motion sensors and 3-axis gyroscopes, It is possible to implement inertial assessment strategies for both velocity and outlook estimation(Davoli et al., 2020).

Inertial sensing is utilized in a variety of ways in the automotive industry, from passenger seatbelts (e.g., airbags and seatbelt pretensioners) to adaptive cruise control (e.g., ESC) to track the motions of a driver while seated.

5.2. Camera Sensors

Video cameras and other imaging-based solutions, are given a certain amount of weight when it comes to stationary and inconspicuous technology that can be used to monitor driver behaviour. To this end, it is commonly recognized that lighting and weather systems have a significant impact on the performance of CV strategies built on top of visual flows from cameras.

5.3. Issues with Sensing Technologies

The fundamental limitation of vision-based and vehicle-based techniques is lighting, which has heterogeneous belongings on both driver mobility tracking and biological signal detection using cameras. Infrared cameras should be used to overcome this constraint because regular cameras perform poorly at night.

Furthermore, the bulk of the approaches presented are only tested in simulation scenarios, which has a detrimental impact on the reliability of this type of research, as "valuable" data may only be gathered after the driver has been in a risky situation.

Figure 5. Methodology

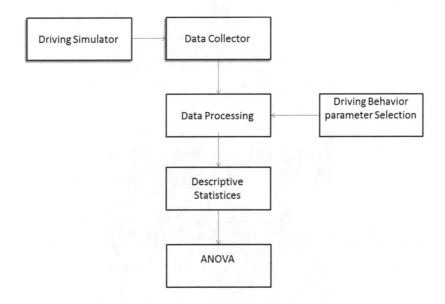

It should be emphasized that bespoke driving scenarios were created using the modules of each specific tool. We employed a deep dnn based on rnn, specifically LSTMs, to construct a prediction algorithm for behavior prediction see Figure 3. LSTMs (Almeida & Azkune, 2018) are universal in the sense that, given enough network units, they can potentially analyze whatever a computer can. These networks are especially well adapted to modeling problems when temporal relationships are important and the time intervals between occurrences are unknown. The ability of LSTMs to prepare for linear data structures has also been demonstrated. The input module, forecast system module, and forecasting module are the three sections of the suggested architecture's LSTMs (Almeida & Azkune, 2018; Ding et al., 2022).

Following the Weight matrix is the forecasting module, which uses the LSTMs' pattern models to forecast the next action. In order to tackle the Recurrent Neural Network's long-term reliance problem (RNN) (Ding et al., 2022; Li et al., 2022) proposed an improved RNN LSTM is a type of RNN that employs a unique gate structure to effectively tackle the problem of slope explosion and disappearance.

Figure 6. Modeling behavior with a (LSTM).

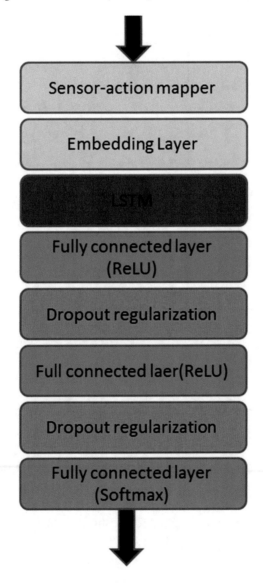

In realm of deep learning (Nimala et al., 2018; Sujatha & Nimala, 2022), CNN is among the most necessary to prioritize. CNN is a feed - forward neural network with a convolutional form that can extract features from data. An encoder, a convolutional layer, a max - pooling, a convolution layer, and an output layer make up the fundamental structure. The convolutional layer uses local connections to conduct deep segmentation method on the original data, while also lowering the original data dimension (Li et al., 2022). The local features can be combined with categorization in the convolutional layer or the max - pooling in the fully connected layer.

Figure 7. Basic Network Structure of the Convolutional Neural Network.

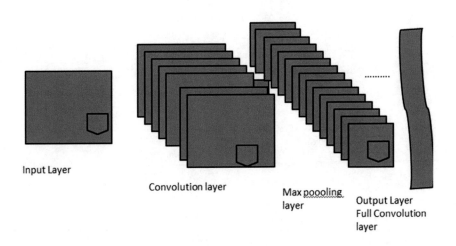

Input Layer

Convolution layer

Max poooling layer

Output Layer
Full Convolution layer

In order to extract features and handle dynamic timing information, CNN models and long short-term memory networks were used. This work uses CNN and LSTM to build a recognition model for vehicle aberrant driving behavior (Li et al., 2022) (Xiao et al., 2022) and see Figure 7 Basic Network structure of the CNN Model. The SVM is a binary data segmentation supervised learning linear classifier (Ding et al., 2022). SVM's purpose is to find the optimal separation hyper plane such that positive and negative observations are correctly detected and the separation between them is as little as possible.

6 USE CASE ON DRIVEN BEHAVIOR RECOGNITION

The flow diagrams from system analysis are transferred onto the network equipment in system design, where the programmer will operate. Figure 8 contain UML Face recognition (Bai et al., 2018; Wrobel, 2018) and figure 9 consists of system architecture.

Maintaining a use case-based perspective and performing a risk analysis for each use case is beneficial for safety-related systems. One benefit of a use case-based approach is that it allows systems with safety-related and non-safety-related system pieces to be treated in the same way.

Figure 8. System Architecture

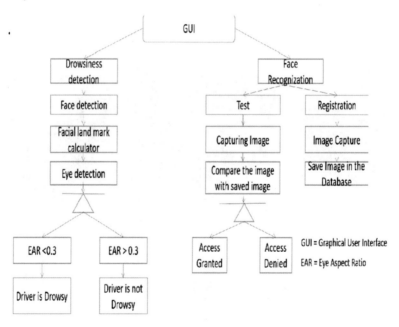

The proposed use case-based technique allows determining either use cases with multiple stages of safety integrity are running separately or if they interact.

7 DATA COLLECTION

Our findings come from a multiple field study in which volunteers supplied a variety of factual sensory data during common trips().

Figure 9. UML diagram for drowsiness detection

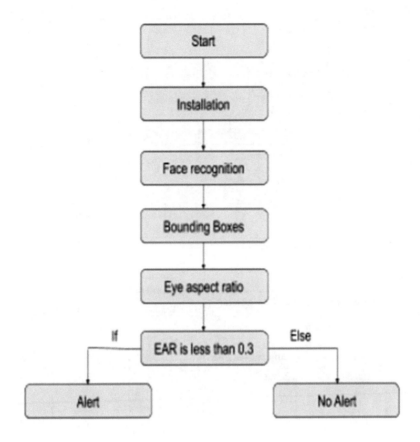

Dataset

The IRTAD database includes crash and exposure data from 32 nations that has been verified and updated: Since 1970, the International Road Traffic and Accident Database (IRTAD) has compiled safety and traffic data by country and year. In the IRTAD countries, all data is acquired directly from appropriate national data providers.

The UK government collects and publishes a lot of data about traffic accidents all around the country (typically once a year). Localities, weather conditions, vehicle kinds, body count, and driving motions are among the topics included in this data. it an interesting and extensive dataset for analysis and research. Figure 12 given dataset available information over internet with year and category.

Figure 10. Activity Illustration

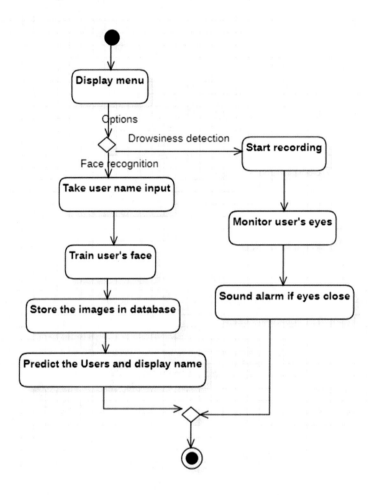

Figure 11. Road Safety Dataset details

Data links

Link to the data	Format	File added	Data preview
Road Safety Data - Accidents 2021 - Provisional Mid Year Unvalidated Data	.CSV	25 November 2021	Not available
Road Safety Data - Vehicles 2021 - Provisional Mid Year Unvalidated Data	.CSV	25 November 2021	Not available
Road Safety Data - Casualties 2021 - Provisional Mid Year Unvalidated Data	.CSV	25 November 2021	Not available
Road Safety - E-Scooter 2021 - Provisional Mid Year Unvalidated Data	.CSV	25 November 2021	Not available
Road Safety Data - Casualties 2020	.CSV	16 October 2021	Not available

Participants

In a company with over 1,000 employees, we used an internal call to recruit nine participants. Our selection was based on the premise of attracting regular travelers. They were selected to represent a wide range of people (). There were two single people and eight married people among the participants. Trying to make phone calls, wearing headphones or the radio, and conversing with other passengers in the car were among the favored activities while driving.

Eye Aspect Ratio (EAR)

The relation of the length and thickness of the eyes is known as EAR. The length and width of the eyes are determined by averaging the two lateral and vertical lines that run throughout the eyes, as shown. The length and width of the eyes are determined by averaging the two lateral and vertical lines that run throughout the eyes, as shown.

Figure 12. Eye Aspect Ratio (EAR)

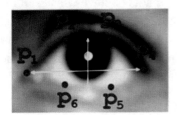

$$EAR = \frac{\|p_2 - p_6\| + \|p_3 - p_5\|}{2\|p_1 - p_4\|}$$

8 SUGGESTIONS AND FUTURE DIRECTIONS

System is potential to dispute and best bit various references and forthcoming possibilities that might be explored framework of the suggested ADAS-oriented design for DCS valuation. The concept of "driver satisfaction" is one component that is recommended to be considered when it comes to ADAS systems. Another suggestion for the forthcoming is to growth the quantity and variety of passengers associated with figures collecting. Other factors to consider in incidents involving motorist sentiment identification via HMIs and ADAS include the subsequent. i) in instruction to offer extra tailored involved facilities, to include driver-related private information and preferences ii) to increase the number and worth of the datasets collected, with the goal of establishing an uniform standard data structure for various applications.

Finally, if Vehicle-to-Everything (V2X) tools are further to (IoT-assisted) enriched ADAS, joint modeling approaches based on numerous vehicles exchanging their on the train acquired data and seeking to improve their center reproductions may be advantageous. As a result, ADAS will be even more powerful systems that assist drivers during their journeys while also enhancing driving quality estimation. Even for emotional state recognition, this can be advantageous because aggregated statistics input might be used by a consortium of trucks to enhance the worth of the identification, while the driver would be ignorant of this smooth handling throughout his or her regular weekly shop.

9 DISCUSSIONS

The driver's handling and assessment of an aid system for friendly driving when joining and turning left were the focus of the study. The traffic scenarios were the same for all of the HMI variations. As a result, there were gaps in the Benchmark and Sensor conditions due to the nearby traffic. As a result, the traffic flow and behavior of the oncoming vehicles had no bearing on the HMI versions' evaluation and iii) to incorporate external context information into models that generate estimates of the driver's state. In compared to traditional manual driving, it is possible to construct more advanced ADASs, especially DMSs, that take into consideration the cognitive and emotive states of the human driver to enable various levels of engagement and automation to progress value of driving and refuge for both motorist and travelers.

10 CONCLUSION

The goal of this study was to improve face component-based techniques' recognition abilities. Our affective computing technology improve the road accident prevention, automobile industries have extended their researches to the development of emotion sensing systems and embedding them inside the vehicles to mitigate the road accidents. Extensive tests were undertaken against various state-of-the-art deep networks to justify the capacity of the model. Finally, recommendations and research directions for the future have been made (covering a variety of factors such as driver satisfaction, the quantity and types of drivers involved in data collecting, and legal and cultural factors that influence drivers).

REFERENCES

Agrawal, U., Giripunje, S., & Bajaj, P. (2013). Emotion and gesture recognition with soft computing tool for drivers assistance system in human centered transportation. *Proceedings of the 2013 IEEE International Conference on Systems, Man, and Cybernetics (SMC'13)*, 4612-4616. 10.1109/SMC.2013.785

Almeida, A., & Azkune, G. (2018, February). Predicting Human Behaviour with Recurrent Neural Networks. *Applied Sciences (Basel, Switzerland)*, *8*(2), 305. doi:10.3390/app8020305

Bai, W., Quan, C., & Luo, Z. (2018, February). Uncertainty Flow Facilitates Zero-Shot Multi-Label Learning in Affective Facial Analysis. *Applied Sciences (Basel, Switzerland)*, *8*(2), 300. doi:10.3390/app8020300

Boril, H., Kleinschmidt, T., Boyraz, P., & Hansen, J. H. L. (2010). Impact of cognitive load and frustration on drivers' speech. *The Journal of the Acoustical Society of America*, *127*. Advance online publication. doi:10.1121/1.3385171

Cruz, A. C., & Rinaldi, A. (2017). Video summarization for expression analysis of motor vehicle operators. *Proceedings of the International Conference on Universal Access in Human-Computer Interaction*, 313-323. 10.1007/978-3-319-58706-6_25

Davoli, L., Martalò, M., Cilfone, A., Belli, L., Ferrari, G., Presta, R., Montanari, R., Mengoni, M., Giraldi, L., Amparore, E. G., Botta, M., Drago, I., Carbonara, G., Castellano, A., & Plomp, J. (2020, December). On Driver Behavior Recognition for Increased Safety: A Roadmap. *Safety (Basel, Switzerland)*, *6*(4), 55. doi:10.3390afety6040055

Ding, D., Gebel, K., Phongsavan, P., Bauman, A. E., & Merom, D. (2014, June). Driving: A road to unhealthy lifestyles and poor health outcomes. *PLoS One*, *9*(6), 1–5. doi:10.1371/journal.pone.0094602 PMID:24911017

Ding, H., Ghazilla, R. A. R., Singh, R. S. K., & Wei, L. (2022, February). Deep learning method for risk identification under multiple physiological signals and PAD model. *Microprocessors and Microsystems*, *88*, 104393. doi:10.1016/j.micpro.2021.104393

Eyben, F., Wöllmer, M., & Schuller, B. (2010). OpenSMILE: The Munich versatile and fast open-source audio feature extractor. *Proceedings of ACM Multimedia*, 1459-1462. DOI:10.1145/1873951.1874246

Gouribhatla, R., & Pulugurtha, S. S. (2022, March). Drivers' behavior when driving vehicles with or without advanced driver assistance systems: A driver simulator-based study. *Transportation Research Interdisciplinary Perspectives*, *13*, 100545. doi:10.1016/j.trip.2022.100545

Grimm, M., Kroschel, K., Harris, H., Nass, C., Schuller, B. B., Rigoll, G., & Moosmayr, T. (2007). On the necessity and feasibility of detecting a driver's emotional state while driving. *Affective Computing and Intelligent Interaction*, *4738*, 126–138. doi:10.1007/978-3-540-74889-2_12

Guo, Yüce, & Thiran. (2014). Detecting emotional stress from facial expressions for driving safety. *Proceedings of the IEEE International Conference on Image Processing (ICIP'14)*, 1, 5961-5965. 10.1109/ICIP.2014.7026203

Jeon, M. (2016). Don't cry while you're driving: Sad driving is as bad as angry driving. *International Journal of Human-Computer Interaction*, *32*(10), 777–790. doi:10.1080/10447318.2016.1198524

Jeon, M., Roberts, J., Raman, P., Yim, J.-B., & Walker, B. N. (2011). Participatory design process for an in-vehicle affect detection and regulation system for various drivers. *Proceedings of the 13th International ACM SIGACCESS Conference on Computers and Accessibility*, 271-272. 10.1145/2049536.2049602

Jones, & Jonsson. (2008). Using paralinguistic cues in speech to recognise emotions in older car drivers. In Lecture Notes in Computer Science: Vol. 4868. *Affect and Emotion in Human-Computer Interaction* (pp. 229–240). Springer. doi:10.1007/978-3-540-85099-1_20

Kolli, A., Fasih, A., Al Machot, F., & Kyamakya, K. (2011). Non-intrusive car driver's emotion recognition using thermal camera. In *Proceedings of the 3rd International Workshop on Nonlinear Dynamics and Synchronization (INDS'11) and the 16th International Symposium on Theoretical Electrical Engineering (ISTET'11)*. IEEE. 10.1109/INDS.2011.6024802

Li, H., Han, J., Li, S., Wang, H., Xiang, H., & Wang, X. (2022, January). Abnormal Driving Behavior Recognition Method Based on Smart Phone Sensor and CNN-LSTM. *International Journal of Science and Engineering Applications*, *11*(1), 1–8. doi:10.7753/IJSEA1101.1001

Ma, Z., Mahmoud, M., Robinson, P., Dias, E., & Skrypchuk, L. (2017). Automatic detection of a driver's complex mental states. *Proceedings of the International Conference on Computational Science and Its Applications*, 678-691. 10.1007/978-3-319-62398-6_48

Moriyama, T. (2012). Face analysis of aggressive moods in automobile driving using mutual subspace method. *Proceedings of the 21st International Conference on Pattern Recognition.*

Morley, D., & Parker, C. (2013). *Understanding Computers: Today and Tomorrow.* Cengage Learning.

N. K., & J. R. (2019, March). A Robust User Sentiment Biterm Topic Mixture Model Based on User Aggregation Strategy to Avoid Data Sparsity for Short Text. *Journal of Medical Systems, 43*(4). PMID:30834466

Nimala, K., Jebakumar, R., & Saravanan, M. (2020, July). Sentiment topic sarcasm mixture model to distinguish sarcasm prevalent topics based on the sentiment bearing words in the tweets. *Journal of Ambient Intelligence and Humanized Computing, 12*(6), 6801–6810. doi:10.100712652-020-02315-1

Nimala, K., Magesh, S., & Arasan, R. T. (2018). Hash tag based topic modelling techniques for twitter by tweet aggregation strategy. *Journal of Advanced Research in Dynamical and Control Systems, 3*, 571–578.

Oehl, M., Siebert, F. W., Tews, T.-K., Höger, R., & Pfister, H.-R. (2011). Improving human-machine interaction: A non invasive approach to detect emotions in car drivers. In. Lecture Notes in Computer Science: Vol. 6763. *Human-Computer Interaction: Towards Mobile and Intelligent Interaction Environments* (pp. 577–585). Springer. doi:10.1007/978-3-642-21616-9_65

Paschero, M., Del Vescovo, G., Benucci, L., Rizzi, A., Santello, M., Fabbri, G., & Frattale Mascioli, F. M. (2012). A real time classifier for emotion and stress recognition in a vehicle driver. *Proceedings of the IEEE International Symposium on Industrial Electronics*, 1690-1695. 10.1109/ISIE.2012.6237345

Sarkar, A., Banerjee, A., Singh, P. K., & Sarkar, R. (2021, September). 3D Human Action Recognition: Through the eyes of researchers. *Expert Systems with Applications, 193.*

Schuller, B., Lang, M., & Rigoll, G. (2006). Recognition of spontaneous emotions by speech within automotive environment. *Tagungsband Fortschritte der Akustik (DAGA'06)*, 57-58. http://www.mmk.ei.tum.de/publ/pdf/06/06sch5.pdf

Siebert, F. W., Oehl, M., & Pfister, H.-R. (2010). The measurement of grip-strength in automobiles: A new approach to detect driver's emotions. In W. Karwowski & G. Salvendy (Eds.), *Advances in Human Factors, Ergonomics, and Safety in Manufacturing and Service Industry* (pp. 775–782). CRC Press. doi:10.1201/EBK1439834992-82

Sujatha, R., & Nimala, K. (2022). Text-based Conversation Analysis Techniques on Social Media using Statistical methods. *2022 International Conference on Advances in Computing, Communication and Applied Informatics (ACCAI)*. 10.1109/ACCAI53970.2022.9752562

Tawari, A., & Trivedi, M. M. (2010). Speech emotion analysis: Exploring the role of context. *IEEE Transactions on Multimedia, 12*(6), 502–509. doi:10.1109/TMM.2010.2058095

Underwood, G., Chapman, P., Wright, S., & Crundall, D. (1999). Anger while driving. *Transportation Research Part F: Traffic Psychology and Behaviour, 2*(1), 55–68. doi:10.1016/S1369-8478(99)00006-6

Wrobel, M. (2018, February). Applicability of Emotion Recognition and Induction Methods to Study the Behavior of Programmers. *Applied Sciences (Basel, Switzerland), 8*(3), 323. doi:10.3390/app8030323

Xiao, H., Li, W., Zeng, G., Wu, Y., Xue, J., Zhang, J., Li, C., & Guo, G. (2022, January). On-Road Driver Emotion Recognition Using Facial Expression. *Applied Sciences (Basel, Switzerland), 12*(2), 807. doi:10.3390/app12020807

Zepf, S., Hernandez, J., Schmitt, A., Minker, W., & Picard, R. W. (2021, May). Driver Emotion Recognition for Intelligent Vehicles. *ACM Computing Surveys, 53*(3), 1–30. doi:10.1145/3388790

KEY TERMS AND DEFINITIONS

ADAS: Advanced driver assistance systems (ADAS) are a collection of electronic technologies that help drivers with driving and parking. ADAS improves automotive and road safety by using a safe human-machine interface.

Annotation Tool: An annotation is additional information that is associated with a specific location in a document or other piece of data. It could be a message with a comment or explanation attached. Annotations appear in the margins of book pages on occasion.

Emotion: A conscious mental reaction (such as rage or terror) is a strong feeling that is usually focused on a single object and is usually accompanied by physiological and behavioural changes in the body.

Human-Machine Interface: A user interface or dashboard that connects a machine, system, or device to a human. While the word HMI can theoretically refer to any screen that allows a user to interact with a device, it is most usually associated with industrial processes.

Inertial Measurement Units (IMUs): An inertial measurement unit (IMU) is an electronic device that uses a combination of accelerometers, gyroscopes, and magnetometers to detect and report the specific force, angular rate, and sometimes the orientation of a body.

User-Centered Design (UCD): During each phase of the design process, designers focus on the users and their demands.

Chapter 8

A Comprehensive Overview of Exercises for Reducing Stress Among Students in Engineering Institutions

Nithyananthan V.
SRM Institute of Science and Technology, India

Rajeev Sukumaran
SRM Institute of Science and Technology, India

Christhu Raj
M. Kumarasamy College of Engineering, India

ABSTRACT

The objective of this work is to evaluate the impact of yoga and meditation intervention on engineering students' stress perception, anxiety levels, and mindfulness skills. Adolescents are experiencing greater stress associated with academic performance, extracurricular activities, and worry about the future. Meditation is the practice by which there is constant observation of the mind. It requires you to focus your mind at one point and make your mind still in order to perceive the 'self'. In this chapter, the authors interrogate the impact of different yogasanas and meditation in enabling learners to get rid of mindful stress. The remainder of the chapter is organized as follows: Section 2 explains the causes of stress and its effects. Section 3 presents the different categories of emotion in stress for explaining the several of levels present in it. Section 4 explains yoga and emotional stress reduction, and Section 5 discusses the conclusions.

DOI: 10.4018/978-1-6684-3843-5.ch008

Copyright © 2023, IGI Global. Copying or distributing in print or electronic forms without written permission of IGI Global is prohibited.

I. INTRODUCTION

THE rapidly changing world demands engineers to be highly effective and efficient in their workplace. This makes current generation student has to work in multitasking features with least importance to their individual health. The transformation of students in their grade level from school habitat to college habitat causes emotional, academic and societal related problems. The learners ware completely exposed to a different learning environment, new technique of ideas in teaching, new academic requirements, new type of connection among learners and facilitators. Due to these changes, learners experience different types of stress. Stress is simply the non-specific reaction of the body to any request made to it. Stress will eventually affect their mental and physical health, which obviously results in poor academic records and other consequences. Stress is one of the principle parts of our advanced life, come about because of the quick changes in human existence, so this age is called as the age of stress. For many young adults, college is the best time of life. Stress becomes an important topic in educational institutions such as Arts & Science, Engineering and Medical education. At present, there is much importance shown on measuring Stress among learners in educational institutions. Research data shows that School is one of the main reasons for developing stress among students. Stress further leads to depression, Physiological problems, and Psychological problems etc, Researchers have found that many mental disorders ware traced to trauma, the damage that happens during college days in times of stress. Various forms of stressors such as Teachers Stress, Results Stress, Peer Stress, Time Management Stress, Self-Inflicted Stress, Burden of Academics with a Responsibility of Achievement, Stress on Uncertain Future "Stress" (Merriam-Webster.com, 2018; Warnecke et al., 2011). Inherently stressful and exhausting, Engineering Education is. The crushing data burden leaves a small chance to relax and recreate and often leads to sleep deprivation. The following statistics in Table 1 shows evidence of stress among college students released by UCLA.

Table 1. Learner's Stress Survey Report

No	Stress	Percentage
1	Depression	19 Million Students
2	Anxiety	1 Million Students
3	Eating Disorders	5-10 Million Students
4	Suicide	1.5 Million Students
5	Mental Distress	1 Million Students

College students ware more stressed out than ever before-as they experience academic stress and social stress. Research by Face and Niles has shown that students ware stressed by mental, emotional, physical and family issues and ultimately have an effect on their learning capacity and academic performance. Although it is difficult for many students to deal with stress, some find stress to be an incentive to work harder to get the most out of the situation. Raj and Simpson revealed that stress contributes to physical and mental health disturbances and stress reduction, and that leading a healthy lifestyle remains a major concern for students in educational institutions. Sheedy research shows that college students are prone to stress and transition of learners from school system to college system, being away from the home for first time, pressure of maintaining high academic achievements leads to stress among freshman students. The College Chronic Life Stress Survey has developed by Tubes, which focused on the level of chronic stress in the lives of college students. This scale includes elements that continue to produce tension over time, such as interpersonal, intrapersonal, academic, and environmental, self-esteem issues and money issues. Research data has measured the stress levels in individual, but failed to provide the necessary remedial measures to avoid stress. Many educational institutions have academic counseling to counsel the student has in order to decrease the student's stress. So that they can achieve better and have a good mental health. They have to provide students with reasonable instructing and learning strategies to diminish their scholastic pressure. Research community have taken effort to measure the stress among students but only very few researchers have proposed remedial measures for overcoming stress. The remedial measures are too vague, generic and do not address to reduce stress in all aspects.

II. CAUSES OF STRESS AND ITS EFFECTS

A. Students Stress

The following section outlines the causes and effects of stress among students. Students react differentially to stressors. Some students may become inspired when confronted with such challenges, and some students may panic. Here are the various stressors to which students are subjected during college days, such as; new levels of freedom, prolonged travel times, living among strangers, roommate negotiations and mediation, unfamiliar environments and climates, heavy course loads, exams, financial responsibilities such as tuition, rent, books, fess, academic results, work schedule, social obligations, romantic relationships.

B. Academic Stress

Student's Academic stress is common among college students. The following reasons causes the academic stress among students; Increased work load, new responsibilities, transferred schools, learning levels is not as good as that of other classmates, lower grade than anticipated, disinterested in chosen program, not interested in some courses, missed too many classes, anticipation of graduation, difference of option with course instructor.

C. Parents Stress

Screening and emergency measure for assessing the nurturing framework and recognizing issues that may prompt issues in the youngster's or parent is conduct. Spotlights on three significant areas of stress: kid qualities, parent attributes and situational/segment life stress.

D. Teachers Stress

Teacher stress: 'The outstanding task at hand wasn't which destitute me – it was the adjustment in my school's way of life'

E. Hostel Stress

The inns of my college, Apeejay Stya University in Sohna, have an upsetting purpose behind why the young ladies' lodging has galleries, while the young men's inn doesn't. As per understudies, back when the college was just a designing school, various male hostellers utilized the overhangs as a way to end it all. I actually could not say whether that is 100% verifiable, notwithstanding, this makes me think about the monstrous pressure undergrads wind up managing. As an undergrad, myself, this was very relatable for me, and I chose to talk with my friends and do some examination on some excellent reasons that can cause misery, stress and uneasiness among understudies.

F. Society Stress

Prevailing difficulty is pressure that comes from one's relationship with others and from the social environment, overall. Considering the assessment speculation of feeling, stress arises when an individual surveys a condition as significant and sees that she does not have the resources for adjust or manage the specific situation. An event which outperforms the ability to adjust does not actually have to occur with the ultimate objective for one to experience pressure, as the risk of such an event happening can be sufficient. Investigators describe prevailing burden and social stressors in an unexpected way. Wadman et al. (2011) described prevalent burden as "the impressions of bother or pressure that individuals may understanding in social conditions, and the connected tendency to keep an essential separation from possibly upsetting social situations". Russell et al. (1977), regardless, portrayed social stressors as "states of step by step social positions that are overall idea to be unsafe or undesirable" (Dormann & Zapf, 2004) portrayed social stressors as "a class of characteristics, conditions, scenes, or practices that are related to mental or real strain and that are by somehow social in nature". There are three crucial characterizations of social stressors. Life events ware portrayed as unexpected, genuine life changes that require an individual to change quickly (ex. assault, unexpected injury). Chronic strains ware portrayed as steady events, which require an individual to make varieties all through a sweeping time (ex. discrete, unemployment). Daily issues ware portrayed as minor events that occur, which require variety for the term of the day (ex. dreadful traffic, disagreements). When stress becomes progressing, one experiences energetic, social, and physiological changes that can put one under more genuine peril for developing a mental issue and physical illness.

G. Job and Career Stress

Word related pressing factor is pressure related to one's work. Word related pressing factor consistently begins from unanticipated obligations and squeezing factors that do not agree with a person's data, capacities, or suppositions, controlling one's ability to adjust. Word related pressing factor could increase when workers do not feel maintained by managers or accomplices, or feel like they have little authority over work processes. Since stress results from the multifaceted associations between gigantic courses of action of interrelated components, there are a couple of mental theories and models that address word related stress (Garber, 2017).

H. Gender Related Stress

Gender contrasts in mind science are contrasts in the mental limits and practices of the sexual orientations, and are a result of a many-sided trade of normal, developmental, and social factors (Wood & Eagly, 2002). Differences have been found in a collection of fields, for instance, mental wellbeing, scholarly limits, character, and tendency towards antagonism. Such assortment may be both inalienable and learned and is routinely difficult to perceive (Harry, 2021). Present day research attempts to perceive such differences, and to inspect any ethical concerns raised. Since lead is a result of relationship among nature and backing investigators are enthusiastic about exploring how science and environment interface with convey such differences, yet this is often not possible.

III. EMOTIONS IN STRESS

This section reveals about the various emotions that causes stress. There are many emotions leads to stress such as; anger, anxiety, apathy, confusion, depression, fear, shame, jealous, happiness, envy, happiness etc. In this article, the authors took few emotions such as Anger, Fear, Shame, Jealous, Happiness, Sadness, and Depression that leads to stress among college students (Tavris, 1989).

A. Anger

All emotions leads to specific skills and gifts. The honorable emotion of all emotions is Anger. Outrage defines limits by strolling the border of mind and watching out for singular objectives. *Benefits*: Outrage can be the most loved feelings that help in seeing precisely who the individual and as a part in social gatherings. Outrage gets one of the good feelings, when understudies realize how to work with it. It causes understudies to turn out to be more responsible, connect really and decent with others. Outrage becomes superb and favorable to social feeling when understudies knows and handles it with following inquiry a) Individual should realize what is it, b) Why it shows up, c) how to work with it. Outrage can uphold you in the sound manner. Essentially put outrage is an important and eminent feeling that can improve your life and your connections goodly.

Classification: Soft Anger: Bothered, Apathetic, Bored, Certain Cold, Crabby, Cranky, Critical, Detached, Displeased, Frustrated, Impatient, Indifferent, Irritated, Peeved, Rankled.

Moderate Anger: Bothered, Apathetic, Bored, Certain Cold, Crabby, Cranky, Critical, Detached, Displeased, Frustrated, Impatient, Indifferent, Irritated, Peeved, Rankled.

High Anger: Aggressive, Appalled, Belligerent, Bitter, Contemptuous, Disgusted, Furious, Hateful, Hostile, Irate, Livid, Menacing, Outraged, Ranting, Raving, Seething, Spiteful, Vengeful, Vicious, Vindictive, Violent

Drawbacks: Too much of anger leads to losing much energy and sometimes problems leading to neuropsychological activities. With a lot of outrage, you may define unbending limits and secure yourself and your suppositions that make different lives hopeless. This situation now and again incorporates individual as well.

B. Fear

Dread can make understudies experience unfavorable reactions physiologically (e.g., windedness), psychologically (failure to center or focus, fanatical speculation, replaying in their brains tricky occurrences that happened in past classes), and inwardly (effectively fomented, defeat by extreme apprehension, dissatisfaction, and other negative emotions) (Zahra et al., 2022). Such degrees of dread may bring about wrong class conduct, ineffectively finished or missing tasks, incessant nonappearances, or exiting courses whenever there is any difficult situation.

Benefits: Fear helps individual to orient, change and connect with your instincts and keep individual safe. Fear helps to focuses on the present moment and immediate surroundings. Generally, people think fear makes oneself uncomfortable, but fear eventually leads to qualities like respecting others, humbleness. Fear helps to focus, identify current position in elation about sensing.

Classification: Soft Fear: Abashed, Awkward, Discomfited, Flushed, Flustered, Hesitant, Humble, Reticent, Self-conscious Speechless, Withdrawn.

Moderate Fear: Ashamed, Chagrined, Contrite, Culpable, Embarrassed, Guilty, Humbled, Intimidated, Penitent, Regretful, Remorseful, Reproachful, Rueful, and Sheepish.

High Fear: Belittled, Degraded, Demeaned, Disgraced, Guilt-ridden, Guilt-stricken, Humiliated, Mortified, Ostracized, Self-condemning, Self-flagellating, Shamefaced, Stigmatized.

Drawbacks: Soft fear have its own functions and benefits. Fear disorders leads to direct result of the way we have all learned and devalue reject and dishonor our goals and objectives.

C. Shame

Disgrace is an excruciating inclination that is a blend of disappointment, self-loathing, and disrespect. A decent individual would feel disgrace in the event that they undermined a test or planned something mean for a companion (Karen, 1992). Feeling disgrace — or being embarrassed — is quite possibly the most hopeless sensations of all. At the point when you feel disgrace, you feel like a terrible individual and lament what you did. In case you are attempting to cause another person to feel awful by chastening them, you are disgracing them. Individuals additionally frequently say, "That is a disgrace," when something terrible occurs — which means it is tragic or a pity (Comstock, 1984).

Benefits: Disgrace helps in changing individual conduct and ensures that singular practices do not do any harm, humiliate, destabilize or dehumanize others. Disgrace is an interesting feeling, because a great many people found out about disgrace by being disgraced. The recuperating practice for disgrace is to uncover inauthentic and applied disgrace, to empower realness, suitable and sound disgrace inside people. Disgrace clearly brings expiation, uprightness, sense of pride and conduct change.

Classification: Soft: Alert, Apprehensive, Cautious, Concerned, Confused, Curious, Disconcerted, Disoriented, Disquieted, Doubtful, Edgy, Fidgety, Hesitant, Indecisive, Insecure, Instinctive, Intuitive, Leery, Pensive, Shy, Timid, Uneasy, and Watchful.

Moderate: Afraid, Alarmed, Anxious, Aversive, Distrustful, Fearful, Jumpy, Nervous, Perturbed, Rattled, Shaky, Startled, Suspicious, Unnerved, Unsettled, Wary, and Worried.

High: Loaded up with Fear, Horrified, Panicked, Paralyzed, Petrified, Phobic, Shocked, Terrorized.

Drawbacks: Shame negative effects leads to toxic and incapacitating emotion, which may result in suicides.

D. Jealous

Envy is the inclination of outrage or harshness, which somebody has when they feel that someone else is attempting to take a darling or companion, or a belonging, away from them E Goldman (2012)[13].

Benefits: Jealous and envy are emotional states that makes individual safe and well positioned in social world. Positive aspects of Jealous lead to Commitment, Security, Connection, Loyalty, and Fairness. Psychological reports suggested that jealousy as primitive emotions for modern day people. *Classification:* Soft: Doubting, Distrustful, Insecure, Protective, Suspicious, and Vulnerable. Moderate: Avaricious, Demanding, Desirous, Envious, Jealous, and Threatened. *High:* Voracious, Gluttonous, Grasping, Greedy, Green with Envy, Persistently Jealous, Possessive Resentful.

Drawbacks: Negative effects of jealously leads in identifying, attracting or relating to reliable companions.

E. Happiness

Understudy satisfaction or misery is not our anxiety. That condition of delight typically has more to do with arranged issues of cash, sex, love, dreams, mental security, state of being, etc (in this regard understudies, similar to us, are individuals as well) (Benjamin et al., 2014). On the off chance that we are to acquire and hold a feeling of the earnestness of our undertakings and to set up some equality of regard with other scholastic projects, we need to quit talking, thinking or stressing over what is not our issue to worry about for example understudy bliss. Should the History teacher on the culmination of his seminar on the Holocaust delay to enquire of his understudies' joy? What is important is the thing that understudies learn.

Benefits: Fear helps individual to orient, change and connect with your instincts and keep individual safe. Fear helps to focuses on the present moment and immediate surroundings. Generally, people think fear makes oneself uncomfortable, but fear eventually leads to qualities like respecting others, humbleness. Fear helps you focus yourself, identify where you are in elation to what you are sensing. Fear gives you energy and focus you need to orient to change or novel situation.

Classification: Soft: Amused, Calm, Encouraged, Friendly, Hopeful, Inspired, Jovial, Open, Peaceful, Smiling Upbeat.

Moderate: Cheerful, Contented, Delighted, Excited, Fulfilled, Glad, Gleeful, Gratified, Happy, Healthy Self-esteem, Joyful, Lively, Merry, Optimistic, Playful, Pleased, Proud, Rejuvenated, and Satisfied.

High: Awe-filled, Blissful, Ecstatic, Egocentric, Elated, Enthralled, Euphoric, Exhilarated, Giddy, Jubilant, Manic, Overconfident, Overjoyed, Radiant, Rapturous, Self-aggrandized, And Thrilled.

Drawbacks: Soft fear have its own functions and benefits. Fear disorders leads to direct result of the way we have all learned and devalue reject and dishonor our goals and objectives.

F. Sadness

Trouble might be the staggering state of mind at a memorial service, for instance, or an old man may depict his life's most prominent misery as letting his youth darling move away. Something intriguing about trouble is that its unique significance was "earnestness."

Benefits: Fear helps individual to orient, change and connect with your instincts and keep individual safe. Fear helps to focuses on the present moment and immediate surroundings. Generally, people think fear makes oneself uncomfortable, but fear eventually leads to qualities like respecting others, humbleness. Fear helps you focus yourself, identify where you are in elation to what you are sensing. Fear gives you energy and focus you need to orient to change or novel situation.

Classification: Soft: Contemplative, Disappointed, Disconnected, Distracted, Grounded, Listless, Low, Regretful, Steady, Wistful.

Moderate Fear: Dejected, Discouraged, Dispirited, Down, Downtrodden, Drained, Forlorn, Gloomy, Grieving, Heavy-hearted, Melancholy, Mournful, Sad, Sorrowful, Weepy, and World-weary.

High Fear: Anguished, Bereaved, Bleak, Depressed, Despairing, Despondent, Grief-stricken, Heartbroken, Hopeless, Inconsolable, and Morose.

Drawbacks: Soft fear have its own functions and benefits. Fear disorders leads to direct result of the way we have all learned and devalue reject and dishonor our goals and objectives.

G. Depression

Melancholy can cause your life to appear to be difficult and inconsequential. It can likewise cause life overall to appear to be unfilled and futile. Just excusing these sentiments as 'silly' or a side effect of 'ailment' disregards the way that inquiries regarding the significance of life are significant issues confronting humankind when all has said in done (Solomon, 2008).

Benefits: Fear helps individual to orient, change and connect with your instincts and keep individual safe. Fear helps to focuses on the present moment and immediate surroundings. Generally, people think fear makes oneself uncomfortable, but fear eventually leads to qualities like respecting others, humbleness. Fear helps you focus yourself, identify where you are in elation to what you are sensing. Fear gives you energy and focus you need to orient to change or novel situation.

Classification: Soft Fear: Apathetic, Constantly Irritated, Angry or Enraged, Depressed, Discouraged, Disinterested, Dispirited, Feeling Worthless, Flat, Helpless, Humorless, Impulsive, Indifferent, Isolated, Lethargic, Listless, Melancholy, Pessimistic, Purposeless, Withdrawn, World-weary

Moderate Fear: Bereft, Crushed, Desolate, Despairing, Desperate, Drained, Empty, Fatalistic, Hopeless, Joyless, Miserable, Morbid, Overwhelmed, Passionless, Pleasure less, Sullen.

High Fear: Agonized, Anguished, Bleak, Death seeking, Devastated, Doomed, Gutted, Nihilistic, Numbed, Reckless, Self-destructive, Suicidal, Tormented, Tortured.

Drawbacks: Soft fear have its own functions and benefits. Fear disorders leads to direct result of the way we has all learned, devalue reject, and dishonor our goals and objectives.

IV. YOGA AND EMOTIONAL STRESS REDUCTION

Parameter and Remedial Measures in Yoga

Anger: Yoga practice for outrage has been demonstrated a profoundly powerful method of containing sessions and explosions of resentment regarding a timeframe. Notwithstanding, Yoga treatment, similar to any remaining meds and treatments ought to be controlled for quite a while for ideal outcomes.

Suggested Yoga practice: Surya Namaskar, *Asanas:* Padmasana, Maha Mudhrasana, Savasana, Makkarasana, Jalandra Bandha, Aswini Mudhra. *Pranayama:* Seethkari Pranayama, Kapalabathi kriya, Nadisudhi pranayama. *Relaxation:* Deep relaxation. *Meditation:* Silent Mediation 15 to 20 minutes, Introspection Practice: Neutralization of Anger

Fear: Fear has mainly manifested as the mental fatigue and physical tiredness. In the yogic management, the techniques has focused to relieve mental as well as physical fatigue or tiredness (Hamann & Gordon, 2000). The practices of asanas are highly beneficial to preserve the energy. Unwanted fear, anxiety and fastness of mind and tired physical state ware replaced with happiness and relaxation by yoga practices. The pranayama practices, the sectional breathing and Nadisuddhi pranayama are more important. These practices slow down the rate of breathing and balance the sympathetic and parasympathetic nervous system. The practice of chandrabhedana helps to the dominance of parasympathetic systems. Lot of relaxation practices is highly needed. For this type of people, meditation may be difficult in the initial stages, but in later stages, meditation has practiced.

Suggested Practices: Surya Namaskar, *Asanas:* Ardhachakarasana, Uttkatasana, Ekapada asana, Ardha Mastsyendrasana, Patchimothasana, Uttanapadasana, Halasana, Shavasana, *Kriyas:* Kapalapathi Kriya, *Pranayama:* Sectional breathing, Nadi sudhi, Ujjayi, Brahmair, *Meditation:* Silent meditation is 15 – 20 minutes, Nine center meditation.

Shame: Association with others is fundamental for connections. At the point when we feel either not exactly or better than they feel another person, we were separated from the other individual. We become irate and accusing. Sensations of inadequacy get comfortable. We need to pull our coats over our heads and become undetectable. That is conveyed disgrace.

Suggested Practices: Hand exercises, Nero muscular breathing exercises, Maharasana exercises, Surya Namaskar, *Asanas:* Padmasana, Vajrasana, Chakrasana, Pavanamukthsana, Savasana, *Pranayama:* Kapalapathy, Nadisudhi, *Meditation:* Agana meditation, Thuriya meditation, Silent meditation

Jealous: Feeling or demonstrating a desirous hatred of somebody or their accomplishments, assets, or saw preferences:

Suggested Practices: Surya Namaskar, *Asanas:* Leg exercise, Food reflexology exercise, Eagapadasana, Yogamudhra, Mahamudha, Nero muscular breathing exercises, Maharasana exercises, *Pranayama:* Kapalapathy, Nadisudhi pranayama, *Meditation:* Agana meditation, Silent meditation, Thuriya meditation, Pancha endhirya meditation, Pancha Boodha Navakiraha Meditation, Introspection:

Analysis of Thoughts

Happiness: It gets the transitory good sentiments that go with euphoria, close by a more significant sensation of significance and reason for the duration of regular daily existence—and proposes how these emotions and sensation of importance reinforce one another.

Suggested Practices: Surya Namaskar, *Asanas:* Neuromuscular breathing exercise, Padmasana, Chin Mudhra, Kayakalpha yoga exercise, *Pranayama:* Kapalabathi kriya, Sithkari prayanayama, Nadisudhi pranaayama, *Meditation:* Thuriyatheetha meditation, Nithiyananda meditation, Pranava meditation, Panchenthiriya meditation, *Introspection:* Eradiation of worries, Neutralization of anger, Moralization of desire, Realization of self.

Sadness: Feelings that has portrayed as a torment felt in the heart, or like you will collapse. Can accompany the response of tears. a few people will in general prefer to remain in dim rooms when affected by pity, others look for frantically for somebody to trust in. a portion of the side effects of misery are crying, cutting, detachment, edginess, glaring, indulging, misjudging, and a great many alternate ways you can adapt to it. Eventual outcomes can differ from migraines and doggy appearances to draining wrists and medical clinic visits. The best treatment is chuckling, or a hug from somebody you care profoundly about.

Suggested Practices: Surya Namaskar, *Asanas:* Neuromuscular breathing exercise, Chin mudhra, Padmasana, Deep relaxation, *Meditation:* Agna meditation, Shanthi meditation, *Introspection:* Eradiation of worries, Realization of self

Depression: Point by point guidelines and showing of yoga rehearses proposed to help assuage wretchedness by fortifying physiological capacities, growing outer mindfulness, and assisting with building up a positive mental disposition.

Asanas are significant in adjusting the actual body, the endocrine framework, and on a more unpretentious level, the chakras and prana in the body. Surya namaskara is of extraordinary advantage for all tension states since it attempts to adjust the whole body and endocrine framework. Ardha chakrasana, ushtrasana the trikosana arrangement chakrasana and dhanurasana work on the adrenals the shakti bandha arrangement; spinal bends likewise makes a difference. Pranayama is a critical part of yoga treatment since it attempts to adjust the nadis and charkras and consequently, the actual body. Ujjayi, the 'clairvoyant' breath, brings tranquility, lucidity and quiet. Bhramari, the murmuring breath, is valuable to ease mental pressures and stresses. Nadi shuddhi is particularly helpful because it works straightforwardly on the nadis, sanitizing the pranic framework and bringing the entire body into balance. Since bhastrika rejuvenates the thoughtful sensory system while kapalbhati tones the parasympathetic sensory system, these practices are reciprocal. Individuals with nervousness will profit by utilizing these practices and figuring out which ones are best for general unwinding, and which, for example, nadi shuddhi, are reasonable during a fit of anxiety.

Sugested Practices: Surya Namaskar, *Asanas:* Ardhakati chakranasana, Padha hasthasana, Ardha chakrasana, Ushtrasana, Trikosana, Chakrasana, Dhanurasana, Savasana, *Pranayama:* Yogic Breathig, Bhastrika, Ujjayi, Nadi sudhi, *Meditation:* Shanthi meditation.

V. CONCLUSION

This overview is the solitary unique duplicate, the extent that we might actually know, that has attempted to mix the diverse composition on the relationship of stress and Yoga/Asanas the opposite method of effect. This emerging focus remains rather than the immense number of studies that have just underscored the anxiolytic and upper effects of action. The current assessment reasons that pressing factor and Yoga has connected in a transient manner. Even more unequivocally, the experience of pressing factor impacts Yoga, and the unprecedented many studies show a contrary association between these forms. Overall, stress impedes individuals' undertakings to be even more really powerful, comparably as it unfavorably impacts other prosperity rehearses, for instance, smoking, alcohol, and medicine use. Unusually, fewer assessments suggest a positive relationship between stress and Yoga. While obviously restricting, these data are consistent with theories that anticipate modifies in lead in either course with pressure. The utility of action as an adjusting or stress the board strategy is prominent and may explain this finding. Strength research suggests that a couple of individuals prosper under conditions of pressing factor; in this way, future assessment is required to understand why a couple of individuals are protected to changes in Yoga in spite of pressing factor while others become inactive. Barely any examinations use exhaustive exploratory plans, which would strengthen this region of solicitation. Regardless, open approaching data is of moderate to high type. Data perceiving arbiters of the association between stress and exercise would help with improving the arrangement of mediations centered towards in peril peoples, for instance, more prepared adults. Future careful assessment around there could be guided by a theory of stress and yoga, which is missing at the present time.

REFERENCES

Benjamin, Heffetz, & Kimball. (2014). Beyond Happiness and Satisfaction: Toward Well-being Indces based on Stated Preference. *American Economic Review, 104*(9), 2698-2735.

Comstock (1984). *Traps for the young.* Academic Press.

Dormann, C., & Zapf, D. (2004). Customer-Related Social Stressors and Burnout. *Journal of Occupational Health Psychology, 9*(1), 61–82.

Garber, M. C. (2017). Exercise as a Stress Coping Mechanism in a Pharmacy Student Population. *American Journal of Pharmaceutical Education, 81*(3), 50.

Goldmen, E. (2012). *Living my life* (Vol. 1). Dover Publications, Inc.

Hamann, D. L., & Gordon, D. G. (2000). Burnout An Occupational Hazard. *Music Educators Journal, 87*(3).

Harry, M. (2021). *Americal on Films, Representing Race, Class, Gender, and Sexuality at the Movies* (3rd ed.). Blackwell Publishing Ltd., John Wiley & Sons.

Karen, R. (1992, February). Sharme. *Atlantic Monthly*, 40–70.

Russell, D., & Kane, J. (1977, November). Sulfide Stress Cracking of High-Strength Steels in Laboratory and Oilfield Environments. *Journal of Petroleum Technology, 29*(11), 1483–1488.

Solomon, R. C. (2008). True to our Feelings: What our emotions are really telling us. Oxford University Press.

Stress. (2018). In *Merriam-Webster.com.* https://www. merriam-webster.com

Tavris, C. (1989). Anger: The misunderstood emotion. Simon & Schuster Inc.

Wadman, R., Durkin, K., & Conti-Ramsden, G. (2011, June). Social Stress in Young People with specific language impairment. *Journal of Adolescence, 23*(3), 421–431.

Warnecke, E., Quinn, S., Ogden, K., Towle, N., & Nelson, M. R. (2011). A randomised controlled trial of the effects of mindfulness practice on medical student stress levels. *Medical Education, 45*, 381–388.

Wood, W., & Eagly, A. H. (2002). A cross-cultural analysis of the behavior of women and men: Implications for the origins of sex differences. *Psychological Bulletin, 128*(5), 699–727.

Zahra, Alivi, & Muazzam. (2022). Exam Anxiety among University Students, Jouranal of Management Practices. *Humanaties and Social Science, 6*(4), 19–29.

Chapter 9

Behavioral Diagnosis of Children Utilizing Support Vector Machine for Early Disorder Detection

Arivarasi A.
Vellore Institute of Technology, Chennai, India

Alagiri Govindasamy
PMCGS Private Ltd., India

Sathiya Narayanan S.
Vellore Institute of Technology, Chennai, India

ABSTRACT

Behavioral issues are categorized by means of persistent difficulties faced from the beginning of childhood. The children with this behavioral disorder restrict social communication and show repetitive interest. Some children have challenging behaviors that are beyond their age and identification becomes difficult. These problems can cause temporary stress on a child's health. A lot of children are impacted by behavioral-related issues from birth, and unfortunately, no scientifically backed early detection mechanism is available to identify the stated issues within the first three years. Using AI and data analytics algorithm, the behavioral profile of a child can be analyzed using a key marker to identify behavioral issues later. Cloud-based AI solutions could be used to implement early detection. Machine learning algorithms using SVM have the potential to create the decision boundary for the segregation using possible classes.

DOI: 10.4018/978-1-6684-3843-5.ch009

Copyright © 2023, IGI Global. Copying or distributing in print or electronic forms without written permission of IGI Global is prohibited.

1. INTRODUCTION

Autistic children are diagnosed during their first few years of life have a much better prognosis because treatment can begin while their brains are still growing. Autism Spectrum Disorder (ASD) and Attention-Deficit/Hyperactivity Disorder (ADHD) are the common childhood brain syndromes that often last into adulthood. One of the most common neurodevelopmental diseases is ASD that influences the children with their social interaction, speech and behavior difficulties in patients which may causes irritability, repetitive behavior and attention issues. ASD has been recognized in the 5th edition of the Diagnostic and Statistical Manual of Mental Diseases (DSM-5), as a bigger umbrella diagnostic category that formerly reflected many separate disorders such as Autism, other Pervasive Developmental Disorders and Asperger's syndrome. Current practice guiding principle for the treatment of ADHD includes diagnosis and assessment, ASD include subsequent the Diagnostic and Statistical Manual (DSM) warning sign categories, which focus on ensuring that symptoms present in many settings, for instance school or home (Räikkönen et al., 2020). These training guiding principles emphasize the significance excluding alternative also/otherwise co-occurring diagnoses for instance such as, mood disorders, learning problems and anxiety, these share key characteristics with ADHD for example difficulty concentration or else ASD, more complicating diagnostic estimation. Even with the fact that these practice recommendations have been developed and revised for over two decades, there's really evidence of significant diversity of extent to which they are used in routine clinical treatment for illness detection.

Inconsistencies in the application of these practice strategies devours the potential which lead to under-, over-, and/or misdiagnosis of the illness. When it comes to diagnosing the disease, a large percentage of specialists ignore the International Classification of Diseases (ICD) or DSM norms entirely. Particularly in the absence of some criteria and/or the child's presentation with a different diagnosis, an average of 16.7% of physicians involved in the diagnosis gave an instance patient an ADHD diagnosis. A false positive rate of over 20% was discovered in follow-up analyses among only those who provided a diagnosis (rather than deferring their diagnostic choice due to a lack of information). While exact figures for ADHD and ASD misdiagnosis aren't available, if the findings of this study are reflective of routine clinical practice, around one-fifth of children identified with ASD or ADHD in the general community are presently misdiagnosed (impacting one million children in the US) (Forness et al., 2012). These children cannot receive treatment for other diseases for instance like anxiety disorders or may receive therapies that are unneeded, causing financial hardship and maybe taking away services that are truly needed by children with these disorders. Particularly in adulthood, several challenges to identification include discrepancies in parent and teacher perspectives, the time investment required for consultations, and faking or malingering symptoms of ADHD and ASD. As a result of these limitations, more ideal assessment approaches for psychological diseases, such as neurobiological or cognitive (tasks).

Machine learning is a branch of artificial intelligence that has the potential to improve the role of computer approaches in neuroscience (Michelini et al., 2016). Deep-learning techniques and machine-learning models have been developed to analyze high-dimensional MRI data in order to replicate neural networks that govern the brains of patients with various mental conditions. These investigations have outcomes in the creation of machine-learning approaches to categorize temporal lobe epilepsy, Parkinson's disease, ADHD, Alzheimer's disease, mild cognitive impairment, autism spectrum disorder, schizophrenia, dementia as well as major depression. These statistical algorithms-based machine-learning models are well-suited to complicated issues requiring a non-linear processes or combinatorial explosion of options, whereas several traditional simulation approaches fall short on quality and scalability.

2. BEHAVIORAL DISORDERS IN CHILDREN

An organization named Finnish Care Register for Health Care had performed disease classification and related issues statistically. This healthcare used Tenth Revision (ICD-10) codes and provided treatment for outpatient and inpatient in the hospital by the physician in order to diagnoses secondary and primary disorder (Lindhiem et al., 2015; Szentiványi & Balázs, 2018). According to validation tests, the Finnish Care Register for Health Care has a high level of psychological validity. Several behavioral disorders in children are discussed as below.

2.1 ADHD

Attention Deficit Hyperactivity disorder (ADHD) is the most frequent and difficult neurobehavioral illnesses in children. ADHD has been shown to have a harmful influence on children, the entire community as well as their families. Approximately one-third to one-half of ADHD patients will continue to have symptoms in adulthood. Socioeconomic background, age, race, gender and the severity of symptoms are all characteristics that can influence a child's or adolescent's diagnosis of ADHD. While boys are diagnosed with ADHD at a higher rate than girls, it is also evident that diagnostic rate and thus, correct detection may differ from gender (Pisano et al., 2017). Boys are more different from girls as they are diagnosed with ADHD, because they are more probable to present with mixed ADHD or impulsive/hyperactive, which is thought to be more damaging. Clinicians are more likely to diagnose ADHD in boys than in females because males comparatively have more externalizing symptoms. Girl children are commonly under detected because they are more prone to present with internalizing indications, such as forgetfulness and difficulties sustaining focus, which are less visible.

A full work up for ADHD includes an assessment for common comorbid symptoms or diagnoses such as anxiety, depression, oppositional defiant and conduct disorders, developmental and learning problems, or other neurodevelopmental disorders. ADHD symptoms can be overshadowed by physical, syndromic, or somatic factors. The occurrence of comorbidities and their seriousness, on the other hand, can enhance the chance of treatment utilization, which may be linked to higher rates of accurate diagnosis. Turner's syndrome affects young women (XO) for instance, more prone to ADHD as well as having this disorder can also confirm or disprove an ADHD detection.

Due to the variety of children in different environment, an accurate clinical assessment requires knowledge of the child's past history and demography. Children from various backgrounds or exposures may have disruptive symptoms that resemble ADHD but are caused by something other than ADHD, such as language barrier or war trauma or a resulting from new migration. Children who grew up in homes where English was the primary language were four times more likely to be diagnosed with ADHD than children who grew up in homes where English was not the primary language. It can be difficult to determine the aetiology of behaviors associated with an ADHD diagnosis because children's psychosocial experiences and the factors that lead to behavioral phenotypes vary greatly.

2.2 ODD

Oppositional Defiant Disorder (ODD) is a common and difficult problem in general medical practice. Despite the fact that it is classified as a separate diagnosis, ODD frequently coexists with the Attention Deficit Hyperactivity Disorder (ADHD). However, unlike ADHD, there has been little research into ODD, and there are no established therapeutic guidelines. ODD is a disorder in which a child acts defiantly, uncooperatively, anxiously, and angrily toward adults. This behavior frequently disrupts the child's normal daily functioning, including relationships and activities in their family and at school. It is not uncommon for children, especially those in their "terrible twos" and early adolescence, to defy authority on occasion (Zhang et al., 2021). They may show their disobedience by arguing with adults, such as their parents or instructors, defying them, or talking back to them. When a child's behavior takes more than six months and is out of character for his or her age, it may indicate that the youngster has ODD.

Vindictiveness, Argumentative/defiant and behavior Angry/irritable mood are the three types of symptoms associated with ODD. Children's mental illnesses, like in adults, those are identified based on signs and symptoms that lead to a specific problem. Regardless of whether the child has symptoms, the physician will begins collecting a comprehensive medicinal history and performing a physical analysis. There are currently no specific laboratory tests to diagnose conduct disorder; however, the physician may employ a variety of techniques, such as prescription side effects, blood tests, or to rule out physical illness as the cause of the symptoms. The doctor will also look for signs of ADHD and depression, which are common co-occurring disorders with ODD. Psychiatrists and psychologists use specially designed interview and evaluation methods to assess a child for a mental condition. The provider makes a diagnosis based on the child's symptoms and observations of the child's attitude and behavior. Because children sometimes have difficulty describing their difficulties or understanding their symptoms, the doctor must frequently rely on reports from the child's teachers, parents, and other adults.

2.3 ASD

Autism Spectrum Disorders (ASD) are a collection of severe developmental abnormalities that affect social connections, communication, play, and academic skills, and often result in lifelong disorder. ASD affects as many as 60 children out of every 10,000. Autism is quite challenging to identify in children at early age, who are frequently submitted to analysis later than is necessary. The parents of the autism children report anxiety at an average age of 17–18 months. The recent data reported the first concern at 14–15 months and also some of the cases were reported at 11 months. However, most of the children in rural areas were diagnosed and reported at 4 years or later (Liu et al., 2020). Early detection and intervention, on the other hand, have been shown to result in a significantly improved prognosis which including adaptive functioning, improved language and social relationships, as well as several maladaptive activities that increasing the possibility of studying in public education. Continued developmental surveillance as well as standardized early screening are the critical facilities for early diagnosis. Professional observations are incorporated into decision-making regarding a child's developmental requirements during developmental surveillance, a continuous process undertaken by pediatric practitioners. Using standardized screening techniques, on the other hand, can improve the accuracy of developmental disorder identification in the context of ongoing pediatric monitoring; using less formal, non-validated approaches results in an unacceptable low sensitivity of 20–30%.

2.4 Conduct Disorders

Conduct disorder is a collection of behavioral and emotional problems that typically begin in childhood or adolescence. Children and adolescents find it difficult to follow rules and behave in a socially acceptable manner as a result of the disease. They may engage in violent, destructive, and deceptive behavior that jeopardizes the rights of others. Adults and other children may view them as "bad" or "delinquent," rather than as suffering from a mental disorder. Whether your child suffers from conduct disorder, he or she may appear tough and self-assured. In truth, children with conduct disorder are generally insecure and mistakenly assume that others are threatening or violent toward them. Children with conduct disorder are notorious for being difficult to manage and disrespectful to rules. They act without considering the consequences of their actions. They are also careless about the feelings of others, such as deception, destructive activity, violent behavior or rule breaking, they may have conduct disorder (Asthana & Gupta, 2019).

3. RISK FACTORS IN CHILDREN'S BEHAVIORAL DISORDERS

Behavioral problems are caused by a variety of factors. A combination of physiological and environmental factors is thought to be involved. It's crucial to remember, though, that a child with a behavioral illness can come from any background gender or sex. The following things may have an impact on their growth.

3.1 Gender

Male children are more susceptible than female children to experience behavioral problems. It's unclear whether this is due to biological differences or whether gender norms and expectations have an impact on how masculine children act and develop. Behavioral disorders are far more common in boys than in girls. It's unknown whether the cause is genetic or related to socialization. Girls with ODD, for example, may be more likely to use words rather than acts to demonstrate aggressiveness. This could indicate that the behavior is less visible and so less likely to be diagnosed.

3.2 Temperament

Early temperamental features may have an impact on a child's development and predict future psychopathology. The environment, on the other hand, can affect or interact with an infant's temperament, making them more vulnerable to negative developmental outcomes. Individual variances in reactivity and self-regulation in the domains of affect, activity, and attention are referred to as temperament. Although a specific description remains a point of contention, the general opinion is that children's temperamental features appear early in life, have a strong genetic or neurological base, and are largely stable over time and conditions (Byeon, 2020; Cohen & Flory, 2019). To present, research has mostly focused on the following temperament dimensions: a) anger/frustration, activity level, positive emotionality, fearfulness, sociability, and attentional orienting in infancy and early childhood; b) Effortful control or the ability to suppress a dominant response in order to conduct a subdominant response are examples of early childhood abilities. In the past, infants with a "difficult" temperament style were described as having negative emotional expressions, inadequate adaptation, strong reactivity, and poor emotional control. However, the term "difficult temperament" refers to a wide range of traits that vary not only across research but also across socio-cultural contexts.

3.3 Gestation and Birth

Complicated pregnancies, low birth and weight premature birth can all contribute to a child's later troublesome behavior. Because of its impact on both mother and child health, maternal mental health research is a public health priority. Once thought to be a time of emotional well-being and "protection" against psychiatric problems, it is now widely recognized that various psychiatric disorders, the most frequent of which is depression, are common during pregnancy. Because of its long-term effects on the mental health and wellness of the mother and her child, violence during pregnancy or intimate relationship violence has also been studied. Because of its long-term effects on the mental health and wellness of the mother and her child, violence during pregnancy or intimate relationship violence has also been investigated (Bazmara et al., 2013). Furthermore, parenting is frequently praised, making the pregnant lady or mother feel guilty for having bad feelings.

3.4 Learning Difficulties

Reading and writing difficulties are frequently linked to behavioral issues. It's tough to figure out what causes mental health problems in children. Unlike certain physiological disorders that have a single cause (for instance, the flu is caused by a virus), mental health issues are thought to be caused by a combination of elements that interact in various ways depending on the kid, family, and social situations. Some of the biological, psychological, and social elements that influence children's mental health are depicted in the diagram to the right. Any of these elements can influence a child's mental health in a positive or negative way (Nelson et al., 2007). Self-esteem, for example, might be high or low, and family conditions can be good or bad at times. Writing, reading and math and so on are all tough for children with learning difficulties. They have discovered a unique impediment to typical teaching methods. Educators can use AI to provide more effective instruction to these students. The application of AI gives accurate feedback as well as simplified instructions and educational materials. There is a scarcity of qualified individuals in India who can help people with learning disabilities.

3.5 Brain Development

Young children were also affected by serious mental health issues. Depression, anxiety disorders, posttraumatic stress disorder, attention deficit hyperactivity disorder, conduct disorder, and neurodevelopmental abnormalities such as autism are common in children. Young children, on the other hand, respond to and process emotional and traumatic experiences in very different ways than adults and older children (Crews et al., 2007). As a result, diagnosing a problem in a child is much more difficult than diagnosing a problem in an adult. The interaction of genes and experience has an effect on children's mental health. Genes do not determine a person's fate. As early as childhood, the combination of genetic predispositions and long-term, stress-inducing experiences can lay the groundwork for an unstable foundation for mental health that lasts well into adulthood. Our genetics contain instructions that tell our bodies how to function, but the chemical "signature" of our surroundings can allow or disallow those instructions from being carried out.

3.6 Family Life

Behavioral disorders are more common in dysfunctional families. A child is at greater risk in families where there is poverty, domestic violence, poor parenting abilities, or substance abuse, for example. It is necessary to investigate the web of social and psychosocial factors that surround the relationship between family structure and health outcomes. This could have ramifications for early intervention programs aimed at reducing morbidity and mortality. Only a few studies have looked at family structure as a variable. There was a flood of studies in the 1960s and 1970s, but interest in this field has waned over time. Furthermore, few studies have looked at hospitalization as a variable, with only a few published in the last few years. When there were multiple mental diagnoses and the severity of maltreatment, the risks of psychiatric readmission increased. Throughout the inpatient rotation, it became clear that the vast majority of the children admitted came from dysfunctional families with a history of abuse.

Disruptions in family structure may have a negative impact on the mental health of both children and their parents. Disruptions have varying degrees of severity. Divorce causes more emotional and behavioral problems in families than other types of interruptions, such as a parent's death. The death of a parent, behavioral issues, and divorce have a greater emotional impact on families than other types of disruptions. Many traits in both children and caregivers have been identified as risk factors for child maltreatment. Depression, poverty, youth, substance abuse, and a family history of women being separated from their mothers as children are all risk factors. Male caregivers face the same risks as women, with the presence of an unrelated male partner at home acting as an additional risk factor. Around 30% of children are expected to live with an adoptive father who is not related to the child. According to research, having a stepparent increases the risk of abuse by a staggering factor of 20–40 times when compared to living with single mothers, where the risk was approximately 14 times higher than living in a biologically intact home. Physical and behavioral abnormalities in children, as well as low birth weight, hyperactivity, and aggression, have all been identified as risk factors. Parents who had experienced childhood abuse or interpersonal violence were more likely to be hostile to their own children. Furthermore, researchers were not able to fully analyze and record the various forms of abuse that children who come from a variety of unstable family structures have experienced.

3.7 Intellectual Disabilities

Individuals with intellectual disability (ID) are frequently diagnosed when they have developmental limitations in both intellectual functioning (IQ of 70 or less) and adaptive behavior (conceptual, practical, and social skills). Intellectual disability is predicted to affect 1.04 percent of the population. However, calculating prevalence figures for this demographic is difficult since they often rely on people who are known to specialist care providers, which may leave out a "hidden" population that does not use services, such as individuals with moderate ID. Intellectual impairment (ID), formerly known as mental retardation, is frequently linked to other medical and psychological illnesses such as cerebral palsy, epilepsy, down syndrome, ADHD, autism, fragile X syndrome and other emotional and behavioral disorders. It is not rare for people with ID to have coexisting psychological illnesses. Dual diagnoses have historically been prevalent in the realm of psychiatry, where the study of intellectual disability falls. However, only recently has research on the presence of psychiatric problems in patients with ID begun. When compared to the general population, people with ID have a higher prevalence of psychiatric problems, ranging from 10% to 80%. Other epidemiological studies have found comparable prevalence rates.

4. SIGNS OF A BEHAVIORAL DISORDER

The behavior associated with oppositional defiant disorder causes problems for the child or teen, as well as the people with whom they interact, at school, at home, and in the community. Anger or irritation, argumentative or defiant behavior, and spitefulness are some of the symptoms of ODD. A kid or adolescent with ODD must show four symptoms from each of these categories to at least one person who is not a brother or sister.

4.1 Emotional Symptoms of Behavioral Disorders

Children and adolescents are affected by a variety of emotional and behavioural disorders, including disruptive, depression, anxiety, and pervasive developmental (autism) disorders, which are classified as either internalizing or externalizing difficulties. Individuals, families, and society all bear the negative consequences of childhood behaviour and emotional difficulties, as well as the disorders that accompany them. They are frequently associated with poor academic, vocational, and psychosocial outcomes. Many healthcare providers, particularly pediatricians, should be aware of the diverse treatment, presentation, and prevention options available for common mental health disorders in children and adolescents (Vismara & Rogers, 2010). Anxiety, depression, and post-traumatic stress disorder (PTSD) are all common emotional problems in adolescence. Many children lack the necessary vocabulary and the ability to communicate their emotions clearly, making it difficult for parents or other caregivers to detect them early. Many physicians and caregivers also have difficulty distinguishing between developmentally appropriate emotions (e.g., anxiety, sobbing) and severe and long-lasting emotional distresses that may be considered illnesses. Disordered eating and low self-esteem are frequently linked to chronic medical conditions such as atopic dermatitis, obesity, diabetes, and asthma, all of which have a negative impact on quality of life.

4.2 Physical Symptoms of Behavioral Disorders

Behavioral syndromes are the types of health problems, which are mostly emotional in nature with no physical indications such as fevers, rashes or headaches. Those with behavioral problems are more likely to develop a substance misuse problem, which might present physically as scorched fingertips, bloodshot eyes or trembling.

4.3 Short-Term and Long-Term Effects of a Behavioral Disorder

A behavioral problem can have both short and long-term negative implications in a person's personal and professional life. Persons who act out may suffer consequences such as suspension or expulsion if they fight, bully, or argue with authority figures. Adults' occupations may be lost in the future. Long-term difficult relationships can cause marriages to fall apart, and children may be forced to change schools until they run out of options.

5. DIAGNOSIS OF CHILDREN'S BEHAVIORAL DISORDERS USING AI

Numerous research has recently established the relevance as well as scope of the human brain system, until now many questions about functions and processes remain unanswered. Significant changes in a kid's learning, behavior, or emotional control are described as child mental health concerns, which cause difficulty and anxiety during the day. Numerous children have anxieties and worries, as well as harmful behavior. If the warning signs are interfered and severe with play, home or school actions on a regular basis, a mental illness may be detected in the child. Brain injury and development begin in early childhood, if not prior to birth. Disorders can be hereditary or result from brain injury from exposure to the environment, such as infection, drug addiction, and foetal alcohol spectrum disorder in the mother, or physical brain injury. DBD impacts people throughout their lives, with symptoms ranging from adults as well as high-functioning children to persons with reasonable to severe scholarly disability, as well as several other prevalent symptoms (Lange et al., 2010). A growing number of tests and biological effect discoveries were published throughout the field of science, ranging from artificial intelligence answers to reports of broader computer technology concerns. Furthermore, AI based approaches do not account for entire brain difficulty and unique characteristics, necessitating new approaches to the cognitive modelling problem.

Emotional, social and psychological well-being are all aspects of mental health. Mental illness can cause problems in physical health, daily life as well as interpersonal connections. Severe changes in pupils' emotional, education or attitude super vision that cause suffering are defined as children's mental disorders. Mental health concerns in children are defined as delays or challenges in the development of age-appropriate thinking. Toddlerhood is a special and joyful time in a child's life when his or her cognitive, physical, and emotional development peaks. Children are often fragile and vulnerable to stress due to changes in their tiny bodies and brains. Stressors could be ubiquitous or unintentional as a result of tonight's television viewing as part of the separation anxiety development phase. Some of the reasons why your baby may be worried, as well as some typical indicators and techniques to relax or lessen your anxiety, are listed below. These issues disturb youngsters and impair their capability to operate jointly in their social contexts, families or classrooms.

Figure 1. Artificial intelligence in mental health.

Figure 1 illustrates artificial intelligence in mental health. Delirium, dementia, amnesia and a form of mental condition that largely impairs perceptions, problem-solving skills, understanding and learning are all examples of cognitive disorders. Schizophrenia is characterized by cognitive impairment. Certain brain areas utilized for distinct cognitive capacities don't always work in people with certain affective or schizophrenia syndromes, according to research. Therefore, it illustrates the mental illness impacts brain function, resulting in cognitive issues. Children's brains adapt well to therapy as they develop, and if they are not treated, if they are not diagnosed early, they are at a higher risk of drug misuse and suicide later in life. A standardized clinician interview and parent questionnaire were used to identify all of the children. Genetic and physiological factors that cause brain diseases, sickness, brain trauma and how they proceed before and during early development are all covered under brain disorders. However, some medications and poisons can have an effect on the brain.

The brain is responsible for social behavior, moods, perception, problem-solving, and reasoning. A brain abnormality may impair a child's normal growth. Depending on their age, as well as the type and severity of their brain illness, they may experience some mental health issues that interfere with their activities or learning. In artificial intelligence (AI), machine learning is considered as an essential in digital approaches to mental health to improve diagnosis, prognosis, and therapeutic strategies for mental health. Artificial intelligence is being used in digital interventions, particularly in web applications and smartphone, in order to improve the user, personalize and experience mental health care. Using AI-driven data techniques, create prediction or detection methodologies for psychological health difficulties utilizing new massive data streams. A deep learning algorithm can detect sorrow and worry in the speech patterns of young children, possibly enabling for the speedy and precise diagnosis of diseases that are difficult to detect in adults. Based on patient data, AI-powered pattern recognition and data analytics could precisely/reliably forecast symptoms. AI can monitor signals to detect mental diseases early on and intervene if necessary. AI-powered solutions could provide low-cost counselling to a large number of patients. People without close confidants or a human support system would be able to receive empathetic therapy via software, which was previously unreachable.

6. MACHINE LEARNING ALGORITHMS TO PREDICT BEHAVIOR DISORDERS

ML techniques, which are a branch of statistics and computer science, study the structures capable and algorithms of learning from observable facts i.e., measurable information. Supervised artificial neural networks learning, non-parametric methods, reinforcement learning, semi-parametric methods, Bayesian decision theory, multivariate analysis, parametric, kernel estimators, hidden Markov models, graphical models and statistical tests these are the available strategies in AI. ML algorithms can alter their parameters based on a data set through a learning process. For instance, this dataset contains all of the coherent data needed to accomplish a categorization, prediction or modelling task. Additionally, it is frequently required to split entirely available data into two subsets such as, 1) The learning set is utilized to learn or calculate the learning machine's optimal parameters, 2) After the machine has learned from the previous set, the test set is used to evaluate its performance. ML techniques have been expedited by the fast-rising number of data received via the Internet and the Internet of Things. Many businesses already have data collection capabilities; the challenge now is figuring out how to put them to good use (Danielson et al., 2018). Depending on the applications and goals, the user model that must be employed varies. It's possible that user models are trying to describe as bellow.

1) The mental processes that underpin a user's actions.
2) A distinction among expert skills and user skills.

3) User preferences or behavioral patterns.
4) The user's characteristics.

The early uses of machine learning techniques in BA were focused on the first two categories of models. Recent research efforts have centered on establishing the third type of model and determining user preferences. Finally, there are few applications of machine learning techniques focused towards discovering user characteristics i.e., those connected to the fourth type of the preceding list. Currently, the scientific subject is the most intriguing to investigate and gets the greatest attention. It's critical to differentiate among techniques for modelling the different users and approaches to modelling communities, classes, and groups of users when creating user models. For predicting behavior issues in children, a variety of machine learning methods are applied. Several techniques are described in the following sections.

6.1 Decision Tree (DT)

To label a pixel, the DT employs a multi-stage or sequential approach, as opposed to the classic statistical and neural/connectionist classifiers, which examine all available information concurrently and make a single membership decision for each pixel. Rather than a single, complex decision, the labelling process is viewed as a series of straightforward decisions based on the results of subsequent tests. The branches of the DT are made up of sets of decision sequences with tests at the nodes. The Decision Tree structure is depicted in Figure 2.

Figure 2. Decision Tree structure

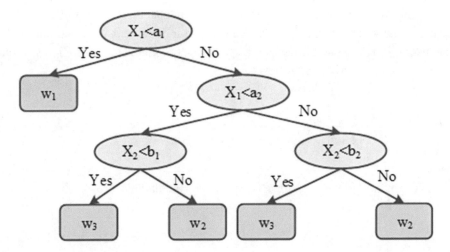

The basis formation of DT is the recurrent division of a collection of training data into progressively homogeneous subgroups based on tests performed on one or more of the feature values. Nodes are used to represent these tests. A single feature is tested in the univariate DT, whereas one or more features are tested simultaneously in the multivariate DT. An allocation mechanism, such as majority voting, is used to allocate labels to terminal (leaf) nodes. DTs were once created by hand, utilizing spectral charts (Srividya et al., 2018).

6.2 Naive Bayes (NB)

One classification technique is the Naive Bayes method. When the class is given, the Bayes theorem-based classifier anticipates that the estimation of a single component will be independent of the estimation of any other element. The most extreme probability is used to select the new class of the new occurrence. They are easy to use and frequently produce accurate categorization results. Their accuracy, however, suffers as a result of their class conditional independence. The Naive Bayes classifier is based on the Bayes theorem with a strong (Naive) independence assumption and works well with large amounts of data (Chen, 2011). The Bayes theorem is used to calculate the likelihood of a document d belonging to class C_j.

$$P\left(C_j|d\right) = \frac{P\left(C_j\right)P(d\,|\,C_j)}{P\left(d\right)} \tag{1}$$

Where $P(C_j)$, $P(d)$, $P(C_j|d)$ and $P(d|C_j)$ are termed the prior, evidence, posterior and probability respectively.

The Naive assumption is that features are independent of conditionality; for example, the appearance of words (features) in a document is not conditionality dependent. If a document possesses a specific set of characteristics $F1,\ldots,Fh$ then equation 3 can be written as follows:

$$P\left(C_j|d\right) = \frac{P\left(C_j\right)\prod_{i=1}^{h}P(F_i\,|\,C_j)}{P\left(F_1,\ldots F_h\right)} \tag{2}$$

An estimation for $P\left(C_j\right)$ *or* $\hat{P}\left(C_j\right)$ can be calculated as:

$$\hat{P}\left(C_j\right) = \frac{A_j}{A} \tag{3}$$

Where A represents the total number of training materials in the category and A_j represents the total number of training papers in category C_j. To classify a new document, Naive Bayes computes posteriors for each class before assigning the document to the class with the highest posterior.

6.3 K-Nearest Neighbor (KNN)

The KNN is an algorithm that becomes more important as computer processing power advances. Computers were not possible prior to the 1960s because they lacked sufficient processing power; but, with the advancement of processors, they have acquired relevance. One of the most readily understood and deployed machine learning algorithms is the KNN method. The distance computation is the foundation of the KNN algorithm (Singh et al., 2017). As a result, the KNN technique must be used on numerical datasets. The Euclidian method is the most widely used distance computation method. The equation for Euclidean calculation is given equation (4).

$$\text{Euclidian}_{i,j} = \sqrt{\sum_{k=1}^{n} \left(x_{ik} - x_{jk} \right)^2} \qquad (4)$$

The KNN algorithm is made up of four phases. The distance between the new data and all data is calculated in the first phase. The distances are then sorted in the second stage. The fewest k values are selected in the third stage, and the class is determined in the last step.

6.4 Logistic Regression (LR)

LR is one of the most widely used statistical models in a variety of analysis models. It defines the relationship between a category variable and one or more dependent variables that can be continuous, binary, or categorical. The approach has the advantage of not requiring variables to have a normal distribution. The independent variables of the LR could be 0 or 1, indicating the presence or absence of landslides. The model output indicates landslide susceptibility on a scale of 0 to 1(Ghazanfar & Prugel-Bennett, 2010). The LR is based on the logistic function Pi, which has been calculated as.

$$P = \frac{\exp(z)}{\left(1 + \exp(z)\right)} \qquad (5)$$

Where P denotes the probability associated with a specific observation and z could be defined as

$$Z = \beta_0 + \beta_1 X_1 + \beta_2 X_2 + \ldots + \beta_n X_n \tag{6}$$

Where, β_0 represent the intercept of the algorithm, n denotes the number of conditioning factors and β_i is the coefficient illustrating the independent variables contribution X_i.

7. DESIGN PROCESS

Machine learning algorithms are used for risk identification, categorization, actual behavior analysis, and risk rating, which can be a difficult task. As a result, many systems use a technique called 'anomaly detection,' which is also referred as 'outlier detection.' The concept is that a user's behavior should be consistent with that of their peers or previous behavior, which is referred to as a baseline. Anomalies are events or observations that are out of the ordinary. This type of anomaly could indicate sabotage, fraud, other malicious intent, data theft, or collusion. When an early deviation is found, the algorithm can either flag it for additional inquiry or, if constructed to do so, compare it to similar incidents in the past. Whenever an early deviation is found, the system can either mark it for future examination or compare it to earlier events. This information could come from a previously run supervised algorithm in which anomalies were classified as "normal" or "abnormal" by a human security analyst, previous training data, or a crowd-sourced knowledgebase, such as numerous customers sharing a threat intelligence database. At last, a risk score is assigned to the threat based on the number of nodes affected, frequency, potential impact, resources involved, and other factors.

7.1 Data Mining

The children behavior related dataset is gathered form National Institute for Health Research (NHS). NHS is Europe's largest national clinical research sponsor, supporting groundbreaking research to enhance people's lives. By providing world-class research facilities, the reserve develops and supports health experts. There are 704 examples in the dataset, each with its own collection of 21 attributes. There are ten questions relating to the behavioral aspects of autism among those 21 attributes that can assist determine if a child has autism or not. Table 1 illustrates the sample attributes in the children behavior related dataset.

Table 1. Sample Features in the Dataset

Feature Id	Features Description
1	Patient age
2	Screening Score
3	Is the user familiar with the screening application?
4	Screening test type
5	Nationality
6	Who is fulfilment the experiment
7	Any member of the family was affected by pervasive developmental problems.
8	The user's country of residence
9	The patient was born with a jaundice condition.
10	Sex

7.2 Preprocessing

Data pre-processing is the process of transforming raw data into a format that is both useable and understandable. The real time data is frequently incomplete and inconsistent; therefore, it contains so many errors and null values. A successful outcome is always the product of good pre-processed data. To deal with incomplete and inconsistent data, several data pre-processing methods are utilized, such as data discretization, outlier detection, handling missing values and data reduction (dimension and numerosity reduction), etc. (Chen, 2011; Tang et al., 2020). The imputation approach was used to solve the problem of missing values in these datasets.

Figure 3. Process flow of behavior analysis

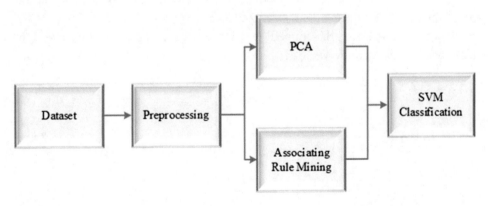

7.3 PCA

PCA is used to minimize the unwanted features in the dataset and it keeps only significant features. Furthermore, the dataset will be reduced in size by maintaining only the most important data, the data set description will be simplified, and the structure of the variation and observation will be investigated. The PCA, which identifies the direction of the data's biggest variance, satisfies all of these goals. The primary components are determined using the principal component (PCs) covariance matrix. The data redundancy is eliminated using PCA in order to accomplish dimension reduction.

Step 1: The dataset is considered as A, which contain n number of features that is represented as, $A=[A_1,A_2,...,A_N]$ and the dataset's dimension is x, and the mean for this dataset is determined here. Equation 1 expresses the mean calculation as below.

$$\sigma = \frac{1}{x}\sum_{k=1}^{x} x_i \tag{7}$$

From equation (7), σ terms the mean calculation, x represents the considered total number of sample and the value of k ranges from 1 to x.

Step 2: The covariance matrix for the sample dataset is constructed using the obtained mean value and the covariance matrix is represented as the below equation (8).

$$\rho = \frac{1}{x}\sum_{k=1}^{x} (a_1 - \sigma)(a_i - \sigma)^T \tag{8}$$

From equation (8) the covariance matrix associated to sample set is denoted as ρ and the calculated mean value is denoted as σ.

Step 3: In this stage, the framed covariance matrix was used to calculate the feature values and feature vectors.

$$\rho = \mathcal{H}.\mathcal{S}\,\mathcal{H} \tag{9}$$

$$\mathcal{S} = diag\left(\omega_1, \omega_2 .....\omega_m\right) \tag{10}$$

$$\mathcal{H} = \left(h_1, h_2 ....h_n\right) \tag{11}$$

From equation (11), the diagonal produced utilizing the covariance matrix with k feature values is denoted by $\mathcal{S}$. In addition, this diagonal matrix is sorted in ascending order. The feature values in the covariance matrix are represented by ωm. The feature matrix that makes up the correlated feature vector hn is represented by $\mathcal{H}$.

Step 4: Using the obtained feature value and feature vector, the cumulative variance contribution rate for the initial v-row element is approximated as follows (Noori et al., 2011).

$$\delta = \sum_{i=1}^{v} \omega_i / \sum_{j=1}^{v} \omega_j \tag{12}$$

In equation (12) is used to estimate value of δ which is may be higher or equal to 0.9. For first v-row items, δ denotes the contribution rate with cumulative variation. The calculated value δ is used to choose the detected problem. The example dataset in question can be accessed by selecting the S value in the v-row element.

Step 5: Finally, lower the dimension of the given sample dataset using the v-row feature vector.

$$\mathcal{B} = \mathcal{H}_v \tag{13}$$

$$\mathcal{N} = \mathcal{B}.A \tag{14}$$

In equation (14), $\mathcal{B}$ denotes feature matrix. Feature vectors are connected to the first v-row items in this feature matrix. The dataset A is changed into $\mathcal{B}$ and x-dimension is minimized into v-dimension. $\mathcal{N}$ Denotes the reduced dimension. The original dataset is reduced using PCA, resulting in effective dimension reduction. By lowering the number of dimensions, categorization accuracy can be increased.

7.4 Associating Rule Mining

Association rule mining in the database is used to uncover new frequency patterns. The goal of association rule mining is to find interesting links and frequency patterns in data sources while reducing correlation between datasets. The two main significant association rules are support (c) and confidence (d). Due to the huge size of the database, users are usually primarily concerned about frequently brought products. The following is a description of how to properly declare association rules mining (García et al., 2011).

Support (s) is known as the percentage of records in a database that comprise all of $M \cup N$ records. For each attribute in the dataset, one attribute is improved. During the scanning procedure, the dataset is browsed over different transactions in any database. The equation (15) shown as follows.

$$Support(MN) = \frac{Support\ sum\ of\ MN}{Whole\ records\ in\ the\ database\ DB} \qquad (15)$$

Confidence(c) is a proportion of the number of transactions; it comprises all records of $M \cup N$ that include M, and it outperforms the confidence ratio threshold-generated association rule $M \Rightarrow N$. The strength of the association's regulations is measured by confidence. The following equation (16) depicts the level of confidence.

$$Confidence(M\ /\ N) = \frac{Support(MN)}{Support(M)} \qquad (16)$$

Designers may select the significant features for effective categorization utilizing this confidence and support in an association rule mining-based rules set. The output attributes from the association rule mining are input into the SVM classifier to forecast the classes. In the following section, we'll go over the SVM classification.

7.5 SVM Classification

SVM (Support Vector Machine) is a supervised learning technique based on statistical learning theory for data search, pattern acceptance, and classification. SVM classification divides data into two groups by creating an N-dimensional hyper plane. The first is the Linear SVM classifier, which divides data points into linear decision boundaries. A hyper plane divides a liner SVM into two sections. The Non-linear SVM classifier, on the other hand, separates the data points utilized to create a non-linear decision boundary. For such tough datasets, non-linear SVM can be used. The SVM Classifier is a classifier that widens the separation between two groups. An SVM's goal is to divide data into two classes by using the training data to separate hyper planes. The "support vectors" are the hyper planes, and the "margin" is the distance between the hyper plane and the nearest support vector. Support vectors with multilayer perceptron and radial-basis function networks can be employed for pattern categorization (Radhamani & Krishnaveni, 2016). Non-linear mapping is used to translate the original training data into a higher dimension. It seeks a linear optimal separation hyperplane within this new dimension. A hyperplane with an adequate nonlinear mapping to a sufficiently high dimension can be used to separate data from two classes. To find these hyperplanes, the SVM employs support vectors and margins. SVM completes the classification task by maximizing the margin and classifying both classes while minimizing classification errors. Although the SVM can be used to solve a variety of optimization problems, the most common is data categorization. The main idea is depicted in Figure 2. The data points are labelled as positive or negative, and the goal is to find a hyperplane that connects them by the shortest distance possible.

Figure 4. SVM classifier

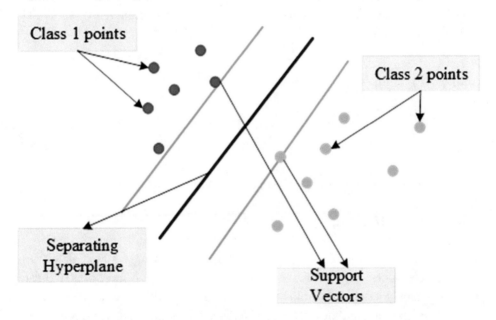

The 2-dimensional example where the data points are linearly separable is shown in Figure 4. For classification and regression, SVM is a linear supervised machine learning technique. It's a program that solves pattern recognition problems. It does not contribute to the issue of overfitting. By defining a decision boundary, SVM separates the classes. The following is the arithmetic for the problem to be solved.

$$\min_{\vec{w},b} \frac{1}{2} w \tag{17}$$

$$st. \ y_i = +1 \rightarrow \vec{w}.\vec{x}_i + b \geq +1 \tag{18}$$

$$y_i = -1 \rightarrow \vec{w}.\vec{x}_i - b \leq -1 \tag{19}$$

$$s.t \quad y_i\left(\vec{w}.\vec{x}_i + b\right) \geq 1, \forall_i \tag{20}$$

Each data point x_i is identified by y_i, which can have a value of +1 or -1 (representing positive or negative respectively). The hyperplane of the solution is as follows:

$$u = \vec{w}.\vec{x} + b \tag{21}$$

The bias is another name for the scalar b. To address this problem, a common way is to use Lagrange theory to convert it to a dual Lagrangian problem. The following is the double issue:

$$\min_\alpha \Psi\left(\vec{\alpha}\right) = \min_\alpha \frac{1}{2} \sum_{i=1}^{N} \sum_{j=1}^{N} y_i y_j \left(x_i.x_j\right) \alpha_i \alpha_j - \sum_{i=1}^{N} \alpha_i \sum_{i=1}^{N} \alpha_i y_i = 0 \tag{22}$$

$$\alpha_i y_i \geq 0, \forall_i \tag{23}$$

The variables α_i are the Lagrangian multipliers for corresponding data point x_i.

7.6 Decision Making

It makes a decision based on the predicted classes. Total scores, for example, range from 15 to 60, with a cut-off number of 30 indicating autism. A score of 30 indicates that the person does not have autism, a score of 30–36 indicates that the person has mild to moderate autism, and a score of 37 indicates that the person has severe autism. This is the score used to determine whether a person has a behavioural disorder or not. Table 2 shows the score for understanding the ADHD disorder.

Table 2. Total Scores Range

ADHD	Nominal illustration
ADHD Measure <=30	No ADHD
ADHD Measure >30 AND ADHD Measure<=50	ADHD mod
ADHD Measure >50	ADHD high

Based on these above scores, the behavior of the children can be easily identified. This can used to find whether a child is normal or up normal in their behavior.

8. ADVANTAGES OF BEHAVIOR ANALYSIS USING MACHINE LEARNING TECHNIQUES

There are numerous advantages to using machine learning techniques. Kernel enables SBMs to define a threshold for distinguishing solvent from insolvent samples, which does not have to be linear or even have the same functional form for all data. This is because its function is non-parametric and operates locally. As a result, they can work with non-monotone functional ratios like the score and likelihood of default, or ones that are not linearly dependent, without having to operate on each non-monotone variable individually. Machine learning is the automatic analysis of behavior-related data that is too much or too complex to analyze using traditional methods in most circumstances including unsupervised learning, reinforcement learning and supervised learning. Data is used to train machine learning models and improve their effectiveness from training using relevant datasets. Once properly educated, they can spot patterns and make judgments on their own, frequently with the same level of accuracy as people.

Machine learning is also adept at detecting hints in a variety of datasets. A variety of factors, such as network login/logout time, travel history, file transfer behavior, location data, job performance, social media interactions, and so on, can be used to identify someone as a dangerous insider. It can then alert a security analyst, who can conduct additional research. The analyst can then utilize other technologies, such as session recording, to further study the behavior to see if it's malicious or just a natural progression i.e., a user assigned a new project triggering a flurry of activities not performed by the user before. The expert's assessment and judgement can then be given back into the system to improve the accuracy of the detection algorithm. Here are some of the benefits of using machine learning algorithms to detect behavior as below.

Reduced number of false positives:

Machine learning can assist prevent such occurrences by reducing false positives while still providing adequate security coverage. It uses several techniques like Clustering, Self-Organizing Maps, Rule-Based Classification and Decision Tree, etc.

Less supervision:

Machine learning results in automation, which eliminates the essential for physical supervision. When configured, the system is capable to handling the vast majority of discovery and classification tasks, as well as, in some cases, automatically reacting to potentially risky user actions.

Earlier prediction as well as response time:

Using current optimal models and hardware, machine learning can do high-speed risk analysis and anomaly detection in enormous amounts of data. As a result, you'll be able to respond to threats faster and more effectively.

Establish correlation & regression:

A person cannot identify and classify data as quickly or efficiently as machine learning can. It's also stronger at distinguishing signal from noise, making it ideal for extracting unusual user behavior from routine tasks.

Scalability:

Machine learning is particularly suited for large-scale deployments because it can manage vast amounts of data from numerous sources. The more data there is, the better the machine can 'learn.'

9. CONCLUSION/SUMMARY

In this chapter the discussion is about Behavioral Diagnosis of children utilizing Machine learning strategy. In the first section, the introduction about several behavior disorders were explained. After that, several Behavioral Disorders in Children such as, ADHD, ODD, ASD and conduct disorders were explained briefly separately. In addition, Risk Factors in Children's Behavioral Disorders as well as Signs of a Behavioral Disorder also illustrated well. Moreover, Diagnosis of Children's Behavioral Disorders Using AI were indicated, which reveals a condition of the normal development of a child. Furthermore, machine learning algorithm for predict behavior disorders were described and, in this section, several machine learning algorithms were used in the behavior analysis such as, and Logistic Regression (LR), K-Nearest Neighbor (KNN), Decision Tree (DT) and Naive Bayes (NB) were discussed. Then the design process involved in the behavior analysis model utilizing SVM machine learning algorithms had described. In this section contain several stages such as, data mining, Preprocessing, PCA for dimension reduction, Associating Rule Mining for selection of effective features, SVM classification and decision making. In this way, the physicians can easily identify whether a child is affection behavior disorders or not.

10. FUTURE WORK

This study focuses on the prediction of Learning Disabilities in children under the age of three utilizing two classification approaches, Support Vector Machine (SVM), and data analytics algorithms. We can quickly and accurately predict autism in any youngster using one of these two classification approaches. For prediction, SVM-based machine learning is particularly effective. Even though the time it takes to build the model is substantially longer, the SVM produces more accurate results than other approaches such as, and Logistic Regression (LR), K-Nearest Neighbor (KNN), Decision Tree (DT) and Naive Bayes (NB). The SVM may take a long time to compute. As a result of this research, we can establish the advantages and disadvantages of these two classifiers when used in the relevant field. The ability of the SVM technique and decision tree model classifier to discover knowledge behind the behavior disorders identification procedure is demonstrated. In this chapter, we'll look at how to handle a learning disability database and compare the pros and drawbacks of Decision Tree and Support Vector Machine classification methods for predicting learning disabilities in school-aged children. Fuzzy sets and close sets will be the focus of our future research in order to predict behavior disorders.

REFERENCES

Althuwaynee, O. F., Pradhan, B., Park, H. J., & Lee, J. H. (2014). A novel ensemble decision tree-based CHi-squared Automatic Interaction Detection (CHAID) and multivariate logistic regression models in landslide susceptibility mapping. *Landslides*, *11*(6), 1063–1078. doi:10.100710346-014-0466-0

Asthana, P., & Gupta, V. K. (2019). Role of Artificial Intelligence in Dealing with Emotional and Behavioural Disorders. *Journal*, *9*(1). http://www. ijmra. us

Bazmara, M., Movahed, S. V., & Ramadhani, S. (2013). KNN Algorithm for Consulting Behavioral Disorders in Children. *Journal of Basic and Applied Scientific Research*, *3*, 12.

Byeon, H. (2020). Development of a depression in Parkinson's disease prediction model using machine learning. *World Journal of Psychiatry*, *10*(10), 234–244. doi:10.5498/wjp.v10.i10.234 PMID:33134114

Chen, M. Y. (2011). Predicting corporate financial distress based on integration of decision tree classification and logistic regression. *Expert Systems with Applications*, *38*(9), 11261–11272. doi:10.1016/j.eswa.2011.02.173

Cohen, I. L., & Flory, M. J. (2019). Autism spectrum disorder decision tree subgroups predict adaptive behavior and autism severity trajectories in children with ASD. *Journal of Autism and Developmental Disorders*, *49*(4), 1423–1437. doi:10.100710803-018-3830-4 PMID:30511124

Crews, S. D., Bender, H., Cook, C. R., Gresham, F. M., Kern, L., & Vanderwood, M. (2007). Risk and protective factors of emotional and/or behavioral disorders in children and adolescents: A mega-analytic synthesis. *Behavioral Disorders*, *32*(2), 64–77. doi:10.1177/019874290703200201

Danielson, M. L., Bitsko, R. H., Ghandour, R. M., Holbrook, J. R., Kogan, M. D., & Blumberg, S. J. (2018). Prevalence of parent-reported ADHD diagnosis and associated treatment among US children and adolescents, 2016. *Journal of Clinical Child and Adolescent Psychology*, *47*(2), 199–212. doi:10.1080/15374416.2017.1 417860 PMID:29363986

Forness, S. R., Freeman, S. F., Paparella, T., Kauffman, J. M., & Walker, H. M. (2012). Special education implications of point and cumulative prevalence for children with emotional or behavioral disorders. *Journal of Emotional and Behavioral Disorders*, *20*(1), 4–18. doi:10.1177/1063426611401624

García, E., Romero, C., Ventura, S., & De Castro, C. (2011). A collaborative educational association rule mining tool. *The Internet and Higher Education*, *14*(2), 77–88. doi:10.1016/j.iheduc.2010.07.006

Ghazanfar, M., & Prugel-Bennett, A. (2010). *An improved switching hybrid recommender system using Naive Bayes classifier and collaborative filtering.* Academic Press.

Lange, K. W., Reichl, S., Lange, K. M., Tucha, L., & Tucha, O. (2010). The history of attention deficit hyperactivity disorder. *Attention Deficit and Hyperactivity Disorders*, *2*(4), 241–255. doi:10.100712402-010-0045-8 PMID:21258430

Lindhiem, O., Bennett, C. B., Hipwell, A. E., & Pardini, D. A. (2015). Beyond symptom counts for diagnosing oppositional defiant disorder and conduct disorder? *Journal of Abnormal Child Psychology*, *43*(7), 1379–1387. doi:10.100710802-015-0007-x PMID:25788042

Liu, G. D., Li, Y. C., Zhang, W., & Zhang, L. (2020). A brief review of artificial intelligence applications and algorithms for psychiatric disorders. *Engineering*, *6*(4), 462–467. doi:10.1016/j.eng.2019.06.008

Michelini, G., Kitsune, G. L., Cheung, C. H., Brandeis, D., Banaschewski, T., Asherson, P., & Kuntsi, J. (2016). Attention-deficit/hyperactivity disorder remission is linked to better neurophysiological error detection and attention-vigilance processes. *Biological Psychiatry*, *80*(12), 923–932. doi:10.1016/j.biopsych.2016.06.021 PMID:27591125

Nelson, J. R., Stage, S., Duppong-Hurley, K., Synhorst, L., & Epstein, M. H. (2007). Risk factors predictive of the problem behavior of children at risk for emotional and behavioral disorders. *Exceptional Children*, *73*(3), 367–379. doi:10.1177/001440290707300306

Noori, R., Karbassi, A. R., Moghaddamnia, A., Han, D., Zokaei-Ashtiani, M. H., Farokhnia, A., & Gousheh, M. G. (2011). Assessment of input variables determination on the SVM model performance using PCA, Gamma test, and forward selection techniques for monthly stream flow prediction. *Journal of Hydrology (Amsterdam)*, *401*(3-4), 177–189. doi:10.1016/j.jhydrol.2011.02.021

Pisano, S., Muratori, P., Gorga, C., Levantini, V., Iuliano, R., Catone, G., & Masi, G. (2017). Conduct disorders and psychopathy in children and adolescents: Aetiology, clinical presentation and treatment strategies of callous-unemotional traits. *Italian Journal of Pediatrics*, *43*(1), 1–11. doi:10.118613052-017-0404-6 PMID:28931400

Radhamani, E., & Krishnaveni, K. (2016). Diagnosis and Evaluation of ADHD using MLP and SVM Classifiers. *Indian Journal of Science and Technology*, *9*(19), 93853. doi:10.17485/ijst/2016/v9i19/93853

Räikkönen, K., Gissler, M., & Kajantie, E. (2020). Associations between maternal antenatal corticosteroid treatment and mental and behavioral disorders in children. *Journal of the American Medical Association*, *323*(19), 1924–1933. doi:10.1001/jama.2020.3937 PMID:32427304

Singh, A., Halgamuge, M. N., & Lakshmiganthan, R. (2017). *Impact of different data types on classifier performance of random forest, naive bayes, and k-nearest neighbors algorithms*. Academic Press.

Srividya, M., Mohanavalli, S., & Bhalaji, N. (2018). Behavioral modeling for mental health using machine learning algorithms. *Journal of Medical Systems*, *42*(5), 1–12. doi:10.100710916-018-0934-5 PMID:29610979

Szentiványi, D., & Balázs, J. (2018). Quality of life in children and adolescents with symptoms or diagnosis of conduct disorder or oppositional defiant disorder. *Mental Health & Prevention*, *10*, 1–8. doi:10.1016/j.mhp.2018.02.001

Tang, H., Xu, Y., Lin, A., Heidari, A. A., Wang, M., Chen, H., & Li, C. (2020). Predicting green consumption behaviors of students using efficient firefly grey wolf-assisted K-nearest neighbor classifiers. *IEEE Access: Practical Innovations, Open Solutions*, *8*, 35546–35562. doi:10.1109/ACCESS.2020.2973763

Vismara, L. A., & Rogers, S. J. (2010). Behavioral treatments in autism spectrum disorder: What do we know? *Annual Review of Clinical Psychology*, *6*(1), 447–468. doi:10.1146/annurev.clinpsy.121208.131151 PMID:20192785

Zhang, X., Wang, R., Sharma, A., & Gopal, G. (2021). Artificial intelligence in cognitive psychology—Influence of literature based on artificial intelligence on children's mental disorders. *Aggression and Violent Behavior*, 101590. doi:10.1016/j.avb.2021.101590

Chapter 10
An Analysis on Multimodal Framework for Silent Speech Recognition

Ramkumar Narayanaswamy
PSG College of Technology, India

Karthika Renuka D.
(iD) https://orcid.org/0000-0002-6519-4673
PSG College of Technology, India

Geetha S.
Vellore Institute of Technology, Chennai, India

Vidhyapriya R
PSG College of Technology, India

Ashok Kumar L.
(iD) https://orcid.org/0000-0001-5962-2961
PSG College of Technology, India

ABSTRACT

A brain-computer interface (BCI) is a computer-based system that collects, analyses, and converts brain signals into commands that are sent to an output device to perform a desired action. BCI is used as an assistive and adaptive technology to track the brain activity. A silent speech interface (SSI) is a system that enables speech communication when an acoustic signal is unavailable. An SSI creates a digital representation of speech by collecting sensor data from the human articulatory, their neural pathways, or the brain. The data from a single stage is very minimal in order to capture for further processing. Therefore, multiple modalities could be used; a

DOI: 10.4018/978-1-6684-3843-5.ch010

Copyright © 2023, IGI Global. Copying or distributing in print or electronic forms without written permission of IGI Global is prohibited.

more complete representation of the speech production model could be developed. The goal is to detect speech tokens from speech imagery and create a language model. The proposal consists of multiple modalities by taking inputs from various biosignal sensors. The main objective of the proposal is to develop a BCI-based end-to-end continuous speech recognition system.

INTRODUCTION

Speech is one of the natural ways of human communication. Person who is suffering from traumatic injuries or any neurodegenerative disorders will not be able to communicate to the public. The studies have said that approximately around 0.4% of the Europe people have a Speech disorder disease. Approximately 40 million American natives suffer from Communication disorders In human life, the Speech and language impairments creates a large impact, they find it very difficult in every aspects of communication activities.. Furthermore, the health-care professionals interact with speech-disabled people is not much comfortable in interacting with them. The people with communication disorders often have a feeling that they are idle and become more stressed since they are not able to communicate like normal people which leads to clinical depression.

The Brain Computing Interface (BCI) is an emerging field in the health care sector. BCI technologies establish a connection between the human brain and the outside world, eliminating the need for traditional data transmission methods. BCI is a technology that uses neural features to help people regain or improve their abilities. Speech BCI would enable real-time communication using brain correlates of attempted or imagined speech. Speech BCI would enable real-time communication using brain correlates to imagined speech. Neural decoders, feature extraction, and brain recording models are all undergoing new improvements. Automatic Speech Recognition (ASR) and related sciences have been hot topics in recent years, and they offer the basis for the Brain Computing Interface.

Silent Speech Interfaces (SSI) is an alternative to traditional acoustic-based speech interfaces. The idea of the Silent Speech Interface is unavailability of intelligible acoustic signals in the speech activity. In recent years Augmented and Assistive Communication have emerged and Silent Speech Interface is one of the assistive devices which are used to restore the oral communication. Lip reading is one of the most familiar method of recognizing the Silent Speech. A variety of other devices are available in order to capture the speech related bio-signals, they are surface Electromyography (sEMG) and Electroencephalography (EEG). In order to capture the electrical activity of the facial muscles with the help of electrodes the (sEMG) sensor is used. The electroencephalogram (EEG) is used to record neural activity in anatomical regions of the brain that are involved in speech production. Because they enable voice communication without relying on the acoustic signal, SSIs are basically new way to restore communication capabilities to human being with speech problems.

SSIs are a sort of assistive technology that helps people regain their ability to communicate verbally. Acoustic Signals are combined with the biosignals produced by various organs during speech production. These biosignals are created by chemical, electrical, physical, and biological processes that occur during speech production by using the biosignals.

Figure 1. Silent Speech Interface

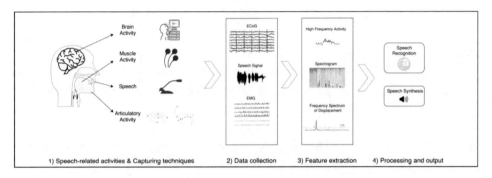

In the above Fig.1, the speech related data is collected from various speech related activities like Brain, muscle, speech and Articulatory. The first step is to collect the data, after the data collection, the feature extraction is performed in order to remove the artifacts, after that it is fed into a deep neural networks and the output is obtained in the form of Speech Synthesis and Speech recognition.

In the next section, the review of literature on the state-of-art methods used for the Silent Speech Interface using different modalities like EEG, EMG and Lip Movement have been discussed.

Literature Review Using Electroencephalogram (EEG) Signal for Silent Speech Interface

Asif, Muhammad, Noreen discussed about the deep learning methods used for the Brain Computing Interface. The author has used the real time EEG dataset and upon request the dataset can be leveraged. The following steps are discussed in this paper, Brain recordings, Preprocessing of the signal, Feature extraction, Implemented using

Machine learning algorithms and Control signals. In order to preprocess the signals Fourier transform or Fourier series are used in order to remove the bad signals in the given input signal, the notch filter is used to remove the irrelevant properties from acquired Brain signal. After preprocessing, the feature extraction is performed for the acquired signals, the Discrete time wavelet transform and continuous time wavelet transform is used to extract the features, in this paper the Short time Fourier transform is used. The reason for applying the wavelet transform is to filter the wavelets of the signal by using the high pass and low pass filters. After extracting the features, the alpha-beta ranges are taken and the features are given as an input to the deep learning network which in turn produced 97.83% of accuracy. The author has finally concluded that, out of the past few years of review, it is found that the Artificial Neural Network (ANN) is used for achieving better accuracy when compared to all other algorithms. The author had concluded that the suggested model can be applied to multiple platforms (Mansoor et al., 2020) (Gu et al., 2020).

Rini and Hema discussed the Correlation based Multiphasal models, the paper combines the EEG signals recorded while speaking, imagining and articulatory movements. The dataset is downloaded from the KaraOne website provided by University of Toronto. After data is collected, using band pass filter, the mean from the dataset is removed. The Handcrafted and CorrNets feature extractions are used for extracting the relevant features. Both the feature extractions are used to extract the Windowed handcrafted feature in the dataset. After the preprocessing and feature extraction, the CorrNet feature extractor is fed into two inputs. The CorrNets have both Canonical correlation analysis (CCA) and Multi-modal auto encoders (MAE). CCA is used to retrieve the common representations and MAE performs self and cross reconstructions of the data. As applied in this dataset it will train in such a way that it maximizes the correlation between the phases in a common place. The author has split into two phases, the training stage has a multi-phasal data and for the testing, the analysis phase data is considered. The main difference that the author had quoted is, the non-activity state from the Kara One dataset is also incorporated for the implementation. The author had used the Gaussian Mixture Model and the Hidden Markov Model for the classification followed by a CorrNet feature extraction method to extract the secondary features from the dataset. The experiment is done using the Kaldi toolkit. The inference of the paper is while using the correlation networks the accuracy is improved than the traditional machine learning and deep learning algorithms. The author has suggested that more models are required in order to obtain better accuracy, Multimodal and Multi Phased data could also be used with a single phase decode approach (Sharon & Murthy, 2020).

Jonathan, Scott et al proposed for evaluating the lightweight mobile EEG devices for speech decoding. The Classification performance for the EEG devices has been discussed in this paper. The various machine learning algorithms for the speech classification and regression with EEG signals has also been discussed. The datasets used in this paper is FEIS (Fourteen Channel EEG with Imagined Speech) dataset. The dataset is divided into three categories training (80%), testing (10%) and validation (10%). The Kara One dataset is also used for classification purposes. In the Kara One dataset 64 channels are used whereas in the FEIS dataset 14 channels are used. The sampling frequency considered for the Kara One dataset is 1000 Hz, in the FEIS dataset 256 Hz is considered. The number of participants involved in the Kara One dataset is 14 and in FEIS, 21 participants are involved. The recording duration for Kara One is 30 to 40 mins and FEIS is around 60 minutes. The 11 syllables are used in Kara One and 16 phonemes are used in the FEIS dataset. The author has used two different classification techniques, Support Vector Machine (SVM) and Convolutional Neural Networks (CNN). On comparing these two classifications for the Kara One dataset, SVM gives good performance compared to CNN. For the regression, in order to predict the vocoder features, two different models, Stacked denoising autoencoder and 7-Layer densely connected neural network is used. These models are used to train the FEIS dataset. Independent Component Analysis (ICA) is used to remove the unwanted noise in the input signal. The authors have concluded that the FEIS dataset can be used for speech synthesize and when it is used with lightweight mobile EEG devices obtains the encoded speech processing same as devices with higher electrodes compared with Kara One dataset. The future work can be, by employing deep neural architectures along with the new optimization techniques on FEIS dataset will provide better decoding performance (Clayton et al., 2019).

Gautam Krishna, Yan et al discussed the Automatic Speech Recognition technique using the Electroencephalography (EEG) signals. The authors demonstrated using two different datasets, the first dataset is based on the Automatic Speech Recognition (ASR) using EEG signals consists of four words and five vowels and the second dataset consists EEG features with larger English vocabulary. The models that are used in this paper are Connectionist Temporal Classification (CTC) Attention based RNN encoder decoder model and RNN transducer model. The time steps allocated for the encoder is the same for all the three models. For the CTC model. The single layer Gated Recurrent unit (GRU) with encoder with 128 hidden units is used. The decoder consists of a dense layer and the softmax activation. The Adam optimizer and the CTC loss function is used here. The model was trained for around 800 epochs. In the second model, the RNN encoder and a decoder is used along with the attention mechanism. The single layer GRU with 512 hidden units is used. The dense and the softmax activation function is used in the second model. The loss function used here is cross entropy and the Adam as the optimizer. The model is trained using the teacher forcing algorithm. The model ran through for around 150 epochs. In the third model, the LSTM with 128 hidden units are used for both encoder and prediction network. The outputs of these networks are fed into a joint network which consists of tanh activation function and again fed into the softmax layer in order to get the prediction possibilities. This model is trained for around 200 epochs. The feature extraction used is IIR band pass filter and notch filter is used for sampling, the notch filter is applied to distill the power line noise. In order to remove other artifacts Independent Component Analysis (ICA) is used. The features such as re root mean square, zero crossing rate and relevant features are extracted. The Kernel Principle Component Analysis (KPCA) is used for dimensionality reduction (Krishna et al., 2020).

The Long Short Term Memory (LSTM) and the Generative model are used in the models to predict listening MFCC from listen EEG (GAN). During the testing period, these two models are used. The RMSE, normalised RMSE, and Mel cepstral distortion (MCD) between output and listen MFCC features from the test set were employed as evaluation measures. The scientists determined that when it comes to speech recognition, the CTC model outperforms the RNN transducer model. When it comes to forecasting MFCC, the LSTM model outperforms the GAN and WGAN models. GAN and WGAN models for speech synthesis are more challenging to train than LSTM models. Future research could take advantage of the enormous amount of EEG corpora available and train the model with more examples. (Krishna et al., 2020).

Damodar Reddy Edla, Shubham et al discussed Using deep learning techniques, the Deceit Identification Test was performed using EEG waves. For recording and analysis, the author employed the Brain Vision Recorder and Analyzer. The band pass filter is used to remove noise from signals during preprocessing. The purpose of the Deceit Identification test is to determine to check whether is person is speaking truth or lying. The analysis is based on real-time data obtained from 30 individuals. 15 were asked to act as an innocent persons and other 15 are asked to be an guilty person. The participants ranged in age from 25 to 35, with 28 men and two women. The EEG signals are used to identify the deception; the P300, which has a positive peak 300 ms to 1000 ms after the stimulus condition, is used in the study. The band pass filter is used to preprocess the signals. For extracting characteristics from signals, the Discrete Wavelet Transform (DWT) is used. The signal is analysed in detail and approximation coefficients using DWT. Following the feature extraction process, the extracted features are fed as input to autoencoder 1, which has ten hidden layers. These features are then reduced and fed as input to autoencoder 2, which has roughly ten layers as well. The softmax layer is used in conjunction with the loss function "cross entropy" to train the features extracted in autoencoders 1 and 2. Because the output is 0 and 1, crossentropy is employed. There are two categories for guilty and innocent in the softmax layer. Multilayer Perceptron, AdaBoost, Quadratic Discriminant Analysis, Decision Tree, Gaussian

Process, Random Forest, Support Vector Machine, and Deep Neural Network are used to evaluate the deception system's performance. In comparison to other classification algorithms, it is assumed that the Random forest, SVM classifier, and Deep neural network produce better results. The proposed approach was also compared to the existing approach by the author. Sensitivity, Specificity, and Accuracy are the evaluation measures employed. The Wavelet Packet Transform (WPT) with Deep Neural Network gives improved accuracy, according to various combinations of algorithms. The main goal of categorization is to distinguish between the true and false classes. The authors came to the conclusion that WPT combined with DNN beats other methods. The accuracy of categorization is improved by using 16 channels, each of which is evaluated and produces a range of results. In the future, the subset of each performing channel, as well as a smaller number of channels that will provide superior performance, will be evaluated.(Dodia et al., 2019).

Literature Review on Using Electromyography (EMG) Signal for Silent Speech Interface

Asif and Krishnan suggested an improvement on the Silent Speech Interface using Surface electromyography (sEMG) signal. The main aim of the paper is to reduce the computational power and also to reduce the channels. The author has implemented using the machine learning algorithms along with the channels. The author aim is to have a limited number of channels and also to reduce the electrodes placed on the face in order to acquire the signals. If the number of electrodes is more then it becomes more invasive to the client. In this paper, the author had used the Decision tree classification algorithm. The more complex algorithms are to be implemented, the computationally less expensive in order to classify the words. The time domain feature is the only feature which is fed as input. The main motivation is to extract the time domain feature from the signals rather than using the frequency domain features. The dataset is collected from EMG UKA Corpus and it consists of 50 words and it is available in Interactive labs in Karlsruhe University. The data is an acoustic data of human speech. The data was collected by placing six channel electrodes and it is placed on the different parts of the articulatory muscles. After the data collection, the features were extracted from the dataset, the author had concentrated on the time domain features rather than frequency domain feature. The reason for choosing the time domain frequency is it reduces the computational issues. After extracting the features, the classification is performed, out of different machine learning algorithms, the author had specifically choosed Decision tree classification algorithm, the reason for choosing this algorithm is, it is less complex algorithm and also it reduces computational burdens. The author had pointed out another important feature of Decision tree algorithm, the datas collected from sEMG had lot of missing values and many number of Outliers. The Decision tree algorithm is less sensitive to Outliers and assumptions on the data are not required. Fifty words were used for the analysis and also to identify the features affecting the word accuracy identification using the sEMG signal. The seven channels have been used, each channel is used to calculate the six time domain features. There were three categorical predictors have been identified, they are, the speaker, the sex of the speaker and the modality of the speech. Totally around 45 input features are fed as an input to the decision tree classification algorithm. In order to extract the time domain features from the dataset, the rectangular window is used. The window should be of length 32 samples and window of one sample. The classification is obtained for all the 50 words along with the seven channels. The result is produced for 100 trails with 95.55% mean word accuracy. In order to calculate the impact of the channel on the classification accuracy, the classification along with the channel combinations are used. The different combinations of the channels have been used and the outcome of the result on these combinations is monitored. Finally, the best channel combination is selected with the accuracy of 95.17%. By reducing the channel, there was a slight variation with the reduction of 0.38% accuracy. The author had concluded that with the help of the time domain features, Decision tree classification algorithm, the number of channels has been reduced and also less computation power has been achieved. The future work narrated by the author is that the Ultrasound and deep learning techniques can be implemented in order to improve the accuracy and also the computational power could also be reduced (Abdullah & Chemmangat, 2020).

Jianhua, Chen and Sunan discussed Electromyography (EMG) signals using the deep belief networks. In this paper the EMG signal is used, the EMG signals mainly used for the action classification. The four time domain feature is extracted from the acquired signals and after the feature extraction, the deep belief network and the generative graphical model are used for classification. The dataset is taken from EMG Physical action dataset from UCI Repository from Essex University. The dataset consists of four subjects, the first three subjects were males and the fourth subjects were female. Totally three males and one female of age category around 25 to 30 yrs are involved in collecting the dataset. Every Subject's EMG consists of 10 normal actions and 10 aggressive actions and around 10,000 samples were collected. There were around eight electrodes used. After the data collection, the next step is extracting the features from the dataset. In this paper, the author has concentrated on four time domain frequencies. The normal actions are categorized like bowing, clapping, standing etc. The aggressive actions are elbowing, front kicking etc. The four time domain features are Zero Crossings, Mean Absolute Difference, Mean Absolute Difference, Mean Absolute Value and Sign Change Slope. The main purpose of selecting these four features is that the Zero Count denotes the number of zero crossings in the signal. The classification is performed using the Deep belief network. The input is the collection of features from the eight channels and output will be of four labels they are, Hand normal and Aggressive, Leg normal and Aggressive, The author had used Restricted Boltzmann Machine. The hidden unit is taken as 50 and number of iterations the author had considered is 300 and only one hidden layer. The dataset is splitted as 50% training set and 50% testing set for the analysis. The performance is calculated and the training and testing samples are divided randomly. The training procedure is repeated for about five times and the testing accuracy is monitored (Dash et al., 2020). The author has also combined the features and the accuracy has been monitored. The final inference of the paper is that the combination of Zero Count, Mean Absolute Difference and Sign Change Slope gave better accuracy. The future work is that more features can be extracted from the EMG signal, including more hidden layers and usage of more subjects like age, health condition etc (Zhang et al., 2019) (Rabbani et al., 2019).

Siyuan, Dantong et al discussed Silent Speech Recognition with the help of Surface Electromyography. This paper analyzes the characteristics of Monosyllabic Chinese characters. Different feature extraction methods and Machine learning classifications are used by the author. The feature extraction methods used by the author are Principal Component Analysis (PCA), Wavelet Packet Transform (WPT) and XGBoost. The algorithms such as Random Forest (RF), Support Vector Machine (SVM) and K-Nearest Neighbor (KNN) are used for the classification. The real time dataset is taken for the analysis. The six Channel electrodes are to be used for the signal capturing. Before collecting the dataset, the skin is cleaned with alcohol and the electrodes are placed on the skin. The subject is to read 10 Chinese characters. The single word was read 20 times and it was completed in one second. Each channel consists of a length of 1s and a sample of 1k. The experiment is conducted for about 3038 samples and these samples are free from noises. The RAW is basically the original sEMG signal, MAV is the average absolute value of the signal and Root Mean Square (RMS). In order to normalize the length of the signals, the Linear Interpolation is used to adjust RAW, RMS and MAV which was reduced from 6000 dimensions to 500 dimensions per channel. The author has discussed the various feature extraction methods and classification algorithms and finally inferred that out of all the feature extraction and classification algorithms, XGBoost is the best feature extraction since it is a tree based structure, the PCA doesn't provide good results. The author has concluded that the Random Forest and Support Vector Machine gave the same result without much difference. Random Forest can be used because the training speed is high and it can be combined in parallel manner, the variance is reduced, the generalization is improved so that the over-fitting can be prevented. The accuracy rate is 72% with the combination of XGBoost with Random Forest algorithm (Ma et al., 2019) (Koctúrová & Juhár, 2018).

Literature Review on Lip Movement for Silent Speech Interface

Jian, Jianzong et al discussed Silent speech recognition using Acoustic signals. The paper focuses on the non-invasive silent speech interface. The non-acoustic signals obtained from the Lip movement are considered for the analysis. In order to capture the signals and listen to their reflections the speaker and microphone of the smartphone are used. The extracted features are fed as an input to the Convolutional Neural Network (CNN) and attention based encoder and decoder network. Since the microphone and speaker are connected to the smart device, this setup can be used for the Lip reading. In this case, sensors are not necessary, only the user needs to move the device near to their mouth while speaking. The data is collected with the help of one Samsung smart phone to collect the lip sentences dataset. All the sentences are in Chinese. There were 10 people involved; the experiment was performed in 8 different locations such as the Laboratory, meeting room etc. Each person will complete five sessions, with each sentence being repeated five times. A total of 13500 samples have been gathered. Data augmentation is used in the training process to expand the dataset. The signal processing pipeline is critical, and the author focused on phase information rather than Doppler Shift. For evaluations, the author had used Word Error Rate (WER). The Word Error Rate under different signal processing is performed. Out of all the mechanisms, phase delta along with the double delta provides the better result. The data augmentation is applied with these mechanisms. (Phase delta+ double delta + data augmentation). The evaluation is divided into three different categories; they are Domain dependent test, Domain independent test, Unseen Sentence test. In the domain dependent test, the domain refers to user and environment, in this test the dataset is segregated as training (70%), validation (10%) and testing (20%), the Word Error Rate of test data is 2.6%. In the Domain independent test, the WER ranges from 3.5% to 12.2%. The average is around 8.4%. In the unseen sentences test, the average WER is 8.1%. The author also compared with Connectionist Temporal Classification (CTC). The author had concluded that since no sensors are required the setup described in this paper can be widely used in Silent voice recognition. The future work could be exploring sequence learning architectures. In the speech recognition tasks, a combination of acoustic signals and the silent signals can be incorporated (Luo et al., 2020).

Hirotaka and Jun discussed the Silent Speech interface by wearing masks and to measure the mouth movement. Mask based Silent Speech Interface is measured the motion around the mouth using two sensors, they are acceleration and singular velocity sensors need to be fit in the mask. With the help of these sensors, motion information of the mouth around 12- dimension is obtained for the analysis. Around 22 states are identified, out of which, 21 are voice commands and 1 is of no speech. The real time datas is collected for the experimenting purposes. Around 21 words are used as an input for Alexa. They are (Alexa, play, music etc.). Three males and one female provided the information. Each user was given four portions, each of which has roughly ten utterances and 22 commands. The user alternated between vocal and silent speaking for each part mask that was removed and worn. Likewise the dataset was collected, another method is to insert the sensors in the skin. Without the mask, the sensors are placed on the chin and right cheeks. As the user speaks, the words uttered are shown in the display and the value of the sensor is monitored for every two seconds later. In total around 880 data has been acquired from each user. After the data collection, it is preprocessed and the data augmentation concept is also used by the author. The data augmentation is done by the factor of 10 depending on how they are choosed. The person is asked to unwear the mask for each time and the location of the mask keeps varies. To overcome this, the data are standardized. Both the sensor datas are standardized by applying the standard deviation. The author has used two neural networks, one is the Lip-interact and the other one is feed forward network. Adam is used for the optimization and Relu as an Activation function with 128 bach size. The collected data is fed into the neural networks, 80% is used for training and 20% is used for the testing set. The input is fed into both the networks, the result has been observed, for command recognition the accuracy was 84.6% and 79.1% for the proposed method. For the speech the accuracy is about 79.9%. The author had concluded that the feed forward network gave better accuracy. SSI with masks is planned to be used for the input of sensitive but personal data, such as information about a person's health (Hiraki & Rekimoto, 2021).

Pingchuan, Brais, Stavros and Maja discussed Lip reading with efficient methods. The paper focuses mainly on training the light weight models and proposed lightweight Architecture. In order to overcome the multiple signals of the same type, the concept called Knowledge Distillation is used. By using the Knowledge Distillation, the author had improved the performance of the model. The dataset is collected from LRW which consists of 1000 speakers from BBC. It had around 50% of utterances of 500 English words. Each has around 29 frames. The preprocessing step the video sequence is taken for the analysis, around 68 facial pictures are tracked using the help of dlib. The Video Sequence of LRW1000 dataset is already cropped. The Adamw is used for the optimization with $\beta 1 = 0.9$, $\beta 2 = 0.999$ and weight decays is around 0.01. The network is trained for about 80 epochs with the learning rate is 0.0003 and a small batch size of 32. The data augmentation concept is used to increase the training samples with the probability of 0.5 and it is shifted horizontally, it is randomly cropped. The lightweight architecture is proposed with the exchange of RESNET-18 architecture into Mobilenet and Shufflenet architectures. The key design here is to replace from depthwise convolution into a normal convolution layer. By doing so, the parameter usage and the number of FLOPS used will be reduced. In order to make a more efficient model, the RESNET-18 has been replaced by the

ShuffleNet v2 Model. It is observed that ShuffleNet is far better than MobileNet, the next step is to replace it from Multi-Scale Temporal Convolution Network (MS-TCN) to Depthwise Separable Temporal Convolutional Network (DS-TCN), this new network will reduce the half of the parameter usage and the Number of FLOPS are still reduced. For the LRW dataset, models can be reduced and made as more lightweight by deducting the heads by 1, by TCN, and also the width of the ShuffleNet to 0.5. The Current model will have less performance compared with the previous defined model 1.5%, and the accuracy will be 78.1%. The LRW 1000, with Shufflenet along with TCN, the performance is reduced by 0.7% compared with the full model. The use of the DS-TCN model does not provide better accuracy. The Shufflenet with 0.5 will have the same performance as the base ShuffleNet v2. The author had inferred that the current model reduces the computational cost for about 8 times compared with the earlier models. The future work is to use the cross modal distillation in the audio visual speech recognition system (Ma et al., 2021).

Minsu, Joanna and Yong discussed the Lip movement using a type of GAN network. In this paper the author had proposed a Visual Context Attentional GAN (VCA-GAN) which can combinely model both local and global lip movements. The proposed model maps the function between Viseme-to-Phoneme. The dataset used in this paper is taken from the GRID Corpus dataset which has fixed grammars from 33 persons. The author had followed three types of settings, they are 1. Constrained Speaker setting 2.Unseen and 3.Multiple speaker settings. All 33 subjects are used for both training and testing. The TCD-TIMIT and LRW datasets are also used for the analysis. For the visual encoding, the 3D Convolutional layer and RESNET -18 are used in this paper. The three generators are used for the implementation, the last two layers are upsampled. The visual encoder is made up of 2 layer bi-GRU along with one linear layer. For the audio encoder, it is constructed with two convolutional layer with two strides and one Residual block. The evaluation metrics concentrated in this paper are STO1, EST01, PESQ and WER. STO1 and ESTO1 are used to calculate the intelligibility of the speech and audio. PESQ is used to calculate the perceptual quality of the speech. WER measures the predicted text from the speech generation. The experimental results discussed in this paper is that the proposed architecture (VCA-GAN) provides better performance compared with the traditional models applied to GRID and TCD-TIMIT dataset except for the STO1 measure. Apart from the above four evaluation metrics, the author had also incorporated the Mean Opinion Score (MOS) test. The subjects are rated from 1 to 5 for the intelligible, naturalness and Synchronization. The proposed architecture generates well-formed speech. The new proposed architecture is applied to the unseen speaker settings, VCA-GAN gives the better performance. In the multi-speaker setting also the proposed architecture provides better results. The proposed Architecture decreases the Computational cost, compared to Seq2Seq based method, VCA-GAN reduces the computational cost. The proposed network takes 25.89ms for producing 3 second speech, whereas Lip2Wav requires around 5 times more than that of the proposed network. The author concluded that VCA-GAN performs better for all the datasets taken for the analysis and it synthesizes the speech from many speakers (Kim et al., 2021).

Gaps Identified from the Literature Review on the three Modalities

- Lack of using Deep learning model for solving speech disabilities and to predict the imagined speech.
- Lack of Using Kinetic motion sensor for capturing the face movements.
- A multimodal fusion framework is required to combine the EEG signal and Kinect-captured muscular action.
- A learning system that can act as an effective control signal for human–computer interactions is lacking.
- The high performance GPU is required to implement the deep learning algorithms.

Proposed Multimodal Framework for Silent Speech Interface

In the below Figure .2 the proposed Multimodal framework is illustrated. The three modalities EEG signals from 64-Channel Neurocap, the sEMG signal from plux and Microsoft Kinetic Camera is used to record facial information and the speech. The collection includes recordings from eight native EP speakers, two of are females and six of them are males, ranging in age from 25 to 35 years. During the recordings, the speakers produced no audible acoustic signal, and only a person who had an experience with silent articulation are considered. Participants were given a 30-minute briefing before each session, which includes rules and speaker preparation. The recording session will be from 40 to 60 minutes, with an average of 3.2GB of data per user, which consists of session metadata, RBG and depth information for a 128x128 pixel square centred at the mouth centre and the coordinates of 100 facial points for each Kinect frame, EMG data from the 5 available channels, two channel wave per prompt containing the Doppler and the synchronisation signal, and a compressed video of the entire session. The speaker was given the cues in a random order. Our aim is to include a variety of participants with varying aged between 15 to 40 years to collect data. All the signals are fed into Convolutional Neural Networks with Adam as an optimizing function and Cross entropy is used here. The Deep Autoencoders are used for training the heterogeneous features produced on both CNN and TCNN. The Deep Autoencoder has 3 encoding and 3 decoding layers and the Mean Squared Error as a cost function.

When data is captured in a single phase, the amount of relevant information available for collection and later processing is reduced. As a result, if a variety of models are combined, in order to obtain the package of speech production model, which would improve speech recognition ability. Because the flaws in one modality can eventually be compensated for by others. The combined performance should be evaluated and compared to that of each modality independently. The proposed innovative SSI uses a Microsoft Kinetic Camera to record facial information, a 64-channel Neuroscan Quick cap to take EEG signals during speaking, and a Plux an sEMG acquisition to obtain the electrical activity from the facial muscles.

Our proposed model framework is composed of Transformers that, instead of recurrent networks like gated recurrent unit (GRU), Long short term memory (LSTM), employ the concept of self-attention, layered layers of self-attention, and positional encoding to learn sequence-to-sequence mapping. Convolutional Neural Network (CNN) and a Deep Autoencoder. The EEG data is pre-processed. Compute features for all 3 modalities, and perform binary classification of phonological categories using a combination of these modalities. The data may be used to learn multimodal relationships, and to develop silent-speech and brain-computer interfaces. The recordings are divided into different states (resting, stimuli, active thinking) Individual networks are trained, and their aggregate outputs are fed through decision level fusion.

Figure 2. 64- Channel Neuroscan Quick Cap

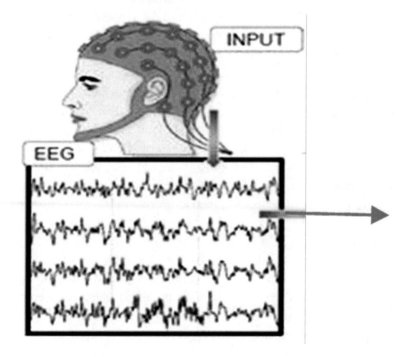

Figure 3. sEMG acquisition system from plux

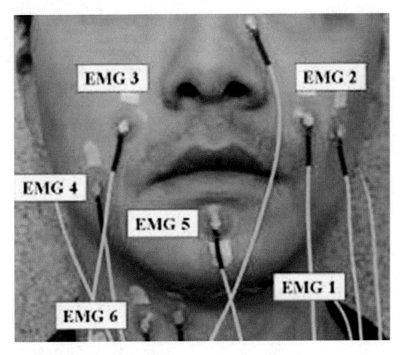

Figure 4. Microsoft Kinetic Camera

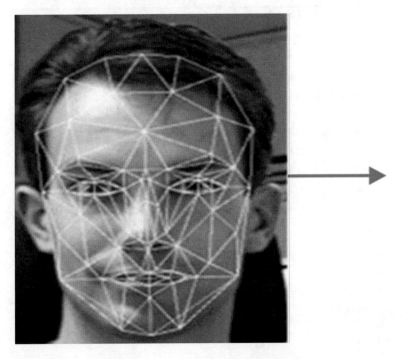

Figure 5.

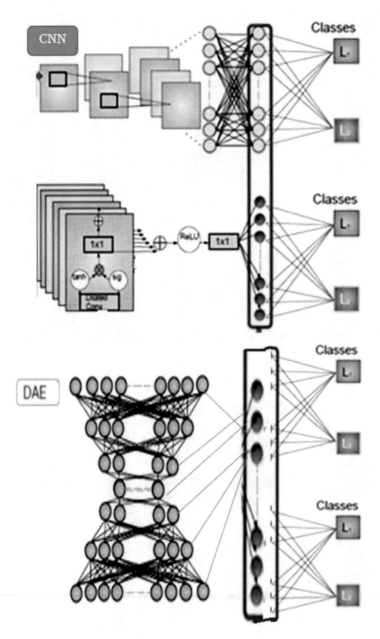

Figures 2, 3, 4, and 5. Multimodal framework for detecting Silent Speech / imagined Speech

CONCLUSION

The Silent Speech Interface is one of the requirements of the Communication disability people. The study so far was done only using the one modality. The traditional approach is concentrated only on the Unimodal approach. The inference of the study is that to propose a Multimodal approach. The proposed Multimodal framework one or more datasets are collected. In our proposed model, the data's are collected from different modalities, after the data collection, preprocessing and feature extraction is carried out, after extracting the relevant features, the fusion concept implemented, after fusing the features, it is fed into classifiers. The Multimodal fusion framework will give better accuracy then the traditional approach.

REFERENCES

Abdullah, A., & Chemmangat, K. (2020). A Computationally Efficient sEMG based Silent Speech Interface using Channel Reduction and Decision Tree based Classification. Elsevier.

Clayton, J., Wellington, S., Valentini-Botinhao, C., & Watts, O. (2019). *Decoding imagined, heard, and spoken speech: classification and regression of EEG using a 14-channel dry-contact mobile headset.* The University of Edinburgh.

Dash, D., Ferrari, P., & Wang, J. (2020). Decoding Imagined and Spoken Phrases from Non-invasive Neural (MEG) Signals. *Frontiers in Neuroscience.*

Dodia, S., Reddy Edla, D., Bablani, A., Ramesh, D., & Kuppili, V. (2019). An Efficient EEG based Deceit Identification Test using Wavelet Packet Transform and Linear Discriminant Analysis. *Journal of Neuroscience Methods, 314,* 1–33. doi:10.1016/j.jneumeth.2019.01.007 PMID:30660481

Gu, X., Cao, Z., Jolfaei, A., Xu, P., Wu, D., Jung, T. P., & Lin, C. T. (2020). EEG-based Brain- Computer Interfaces (BCIs): A Survey of Recent Studies on Signal Sensing Technologies and Computational Intelligence Approaches and their Applications. arXiv preprint arXiv:2001.11337

Hiraki, H., & Rekimoto, J. (2021). SilentMask: Mask-type Silent Speech Interface with Measurement of Mouth Movement. Association for Computing Machinery.

Kim, M., Hong, J., & Man Ro, Y. (2021). Lip to Speech Synthesis with Visual Context Attentional GAN. *35th Conference on Neural Information Processing Systems (NeurIPS 2021),* 1-13.

Koctúrová, M., & Juhár, J. (2018, August). An overview of BCI-based speech recognition methods. In *2018 World Symposium on Digital Intelligence for Systems and Machines (DISA)* (pp. 327-330). IEEE. 10.1109/DISA.2018.8490536

Krishna, G., Han, Y., Tran, C., Carnahan, M., & Tewfik, A. (2020). *State-of-the-art Speech Recognition using EEG and Towards Decoding of Speech Spectrum from EEG.* arxiv: 1908.05743v5.

Luo, J., Wang, J., Cheng, N., Jiang, G., & Xiao, J. (2020). *End – to – End Silent Speech Recognition with Acoustic Sensing.* arxiv: 2011.11315v1.

Ma, P., Martinez, B., Petridis, S., & Pantic, M. (2021). Towards Practical Lipreading with Distilled and Efficient Models. doi:10.1109/ICASSP39728.2021.9415063

Ma, S., Jin, D., Zhang, M., Zhang, B., Wang, Y., Li, G., & Yang, M. (2019). Silent Speech Recognition Based on Surface Electromyography. IEEE.

Mansoor, A., Usman, M., Jamil, N., & Naeem, A. (2020). Deep Learning Algorithm for Brain-Computer Interface. *Scientific Programming*, *2020*, 1–12. doi:10.1155/2020/5762149

Rabbani, Q., Milsap, G., & Crone, N. E. (2019). The potential for a speech brain–computer interface using chronic electrocorticography. *Neurotherapeutics; the Journal of the American Society for Experimental NeuroTherapeutics*, *16*(1), 144–165. doi:10.100713311-018-00692-2 PMID:30617653

Sharon, R., & Murthy, H. (2020). Correlation based Multi-phasal models for improved imagined speech EEG Recognition. doi:10.21437/SMM.2020-5

Zhang, J., Ling, C., & Li, S. (2019). *EMG Signals based Human Action Recognition via Deep Belief Networks.* Elsevier Ltd.

Chapter 11

Voice–Based Image Captioning System for Assisting Visually Impaired People Using Neural Networks

Nivedita M.
Vellore Institute of Technology, Chennai, India

AsnathVictyPhamila Y.
Vellore Institute of Technology, Chennai, India

Umashankar Kumaravelan
Independent Researcher, India

Karthikeyan N.
iD https://orcid.org/0000-0002-6199-5131
Syed Ammal Engineering College, India

ABSTRACT

Many people worldwide have the problem of visual impairment. The authors' idea is to design a novel image captioning model for assisting the blind people by using deep learning-based architecture. Automatic understanding of the image and providing description of that image involves tasks from two complex fields: computer vision and natural language processing. The first task is to correctly identify objects along with their attributes present in the given image, and the next is to connect all the identified objects along with actions and generating the statements, which should be syntactically correct. From the real-time video, the features are extracted using a convolutional neural network (CNN), and the feature vectors are given as input to long short-term memory (LSTM) network to generate the appropriate captions in a natural language (English). The captions can then be converted into audio files, which the visually impaired people can listen. The model is tested on the two standardized image captioning datasets Flickr 8K and MSCOCO and evaluated using BLEU score.

DOI: 10.4018/978-1-6684-3843-5.ch011

Copyright © 2023, IGI Global. Copying or distributing in print or electronic forms without written permission of IGI Global is prohibited.

INTRODUCTION

Image Captioning is the method which involves perceiving a particular scene or image and formulating connections among various objects in a image and assigning a concise depiction or rundown of the image/scene. The field of deep learning has progressively faced development in the architectures and methods used for image captioning. But generally these deep learning models adhere to a guideline structure with not many alterations.

The entire model generally comprises of two sub-models: A Encoder (CNN) for separating highlights from the image, A Decoder (NLP Language Model) for producing the subtitles in light of the information highlights. The Encoder's output is straightforwardly passed to the NLP Model alongside the train subtitles during the training phrase. Alongside this model design, attention models are additionally executed to mirror the visual attention of a genuine human being to capture and leverage features and visual elements when generating a word based on the image.

In this chapter, we shall give a brief into the major components involved in image captioning, a summary of those components and some examples of available image captioning methods, metrics and datasets. We shall then take a look at the proposed system for image captioning for the visually impaired.

Figure 1. General Image Captioning Architecture (Left: Image Feature Extraction Right: Language Model generates caption outputs (y1,y2,..) from input x1,x2,..)

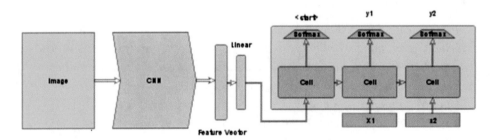

Computer Vision

In Computer Vision, the main element in many algorithms is called a filter. A filter is used to extract a particular type of information from the image. For example, the Sobel and Prewitt filters are used to extract edges. Similarly, we can make algorithms learn filters for colors, shapes and other image features.

Figure 2. Edge Detection Filters (Left-Right: Vertical, Horizontal, Diagonal Filters)

0	1	0
0	1	0
0	1	0

0	0	0
1	1	1
0	0	0

0	0	1
0	1	0
1	0	0

Convolutional Neural Networks

The base architecture model of any Image related Deep Learning Model is CNN. An image is taken as input by a Convolutional Neural Network (ConvNet/CNN) and weights and biases will be assigned to all the identified objects in the image which will be helpful in differentiating one from the other. In the previous section we talked about filters.

The characteristics of the image will be learnt by these ConvNets .So using these CNNs we can extract millions of such features and then pass these features to a Feed Forward Neural Net to classify these images. Since CNNs can easily learn different features quickly, the pre-processing required is minimal when compared to other classification algorithms. Generally CNNs consist of three major layers:

- Convolutional Layer(CL)
- Pooling Layer(PL)
- Fully Connected Layer(FCL)

Convolutional Layer

A CL is the main layer which consists of a set of filters. Each filter is convolved across the input image and the dot product between the filters and the input will be computed and the result will be the filter's 2-dimensional activation map. The network can learn the filters that will be activated when certain type of the feature is detected from the input at some spatial position.

Pooling Layer

Pooling layer will occur in between two convolutional layers. The dimensionality of the input data has to be reduced so that the computing power can be made minimal. The rotational and positional invariant features also will be extracted from the pooling layer which helps in smooth running of the training process. Maximum pooling and average pooling are the two prominent pooling types. It is an important layer for detecting objects from an image.

Fully Connected / Feed Forward Layer

All the high-level features which is represented by the output of the CL will be learnt by the fully connected layer. This layer tries to learn a function which takes in the data in that space. After a stack of CL and PL, the output is then flattened to a 1 - dimensional vector and is passed as input to the FCL to be mapped to the required output label.

Figure 3. A general representation of a CNN (At each layer the input is taken and a set of filter representations are learnt which are then pooled into a smaller set of feature maps)

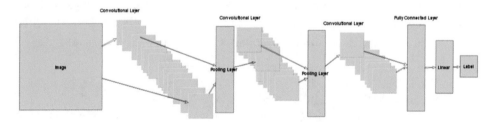

Natural Language Processing (NLP)

NLP involves the subfield of linguistics, artificial intelligence and deep learning which leverages the use of algorithms to analyze, interpret and compute contextual patterns and produce insights and particular features from text and audio data sources.

The most common architectures in natural language processing rely on sequence modelling, where the model tries to compute the outputs based on a sequential understanding of inputs. This involves understanding data and it's correlations between sequences and data sequence understanding across time as well. This is more commonly referred to as spatio-temporal understanding. The most common types of language model components are as follows:

- Recurrent Neural Networks (RNNs)
- Long – Short Term Memory (LSTMs)

Recurrent Neural Networks

RNNs are the building blocks of sequence modelling in deep learning. The appropriate term for these blocks is called as an RNN Cell. The property of RNN cells is that they are able to take an input and compute an output while simultaneously being able to update their internal state using the weights learnt from the computation of previous cell outputs.

Figure 4. An internal representation of a Recurrent Neural Cell (The input Xt is taken and combined with the state of the previous recurrent cell. A non-linearity on this provides us with the output and the weight matrix is sent as next state representation)

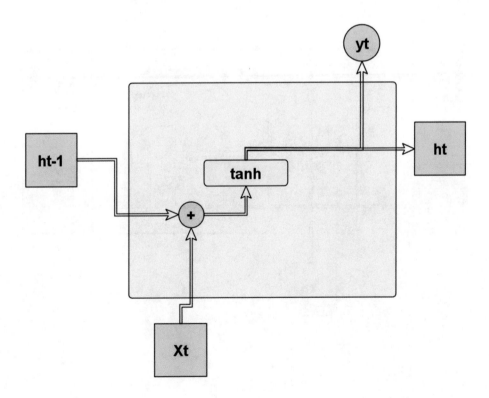

Long – Short Term Memory

LSTM cells are slightly an advanced version of RNN cells. They have three different gates like Input Output and Forget Gate. Similar to the RNN the input and output gates perform computations on the input and generate the output. The forget gate on the other hand, takes in the input of the current cell and output of the previous cell and computes a weighted computation followed by a non-linearity computation such as tanh activation to determine whether the previous input information is required for the next input or to forget the information.

Figure 5. The Long-Short Term Memory Cell. (Left: Forget Gate, Mid: input and cell update gate, Right: Output Gate) The input and the previous cell states are taken and non-linearity such as sigmoid/tanh is computed. These represent the different functions such as the cell update, forget gate, output gate.

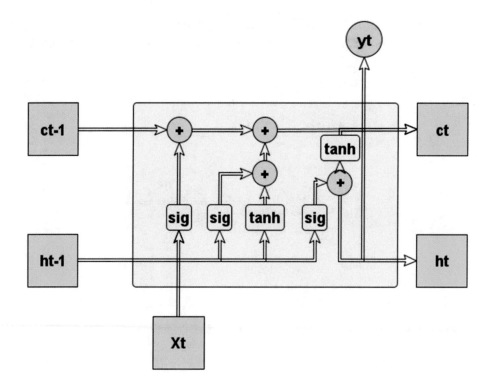

Image Captioning System

From the introduction, we can regard Image Captioning as a sequence learning problem, where it converts a sequence of image features to a word sequences. For this specific goal, we need to process both the image into an interpretable set of features and convert the ground truth captions or image descriptions into a sequence of tokens. This preprocessing of both image and caption data involves the following sequence of operations:

- Cleaning Descriptions
- Vocabulary Construction
- Tokenization of Descriptions
- Sequence Generation
- Image Feature Extraction

Cleaning Descriptions

This includes removal of punctuation, single character occurences, numbers, and conversion to lowercase and other sentence level modifications (ASCII, Emoji, Symbol conversion/removal).

Vocabulary Construction

This step involves compiling a corpus of words which are present in the entire dataset. This vocabulary set is used in the tokenization process to convert words to a sequence of interpretable features.

Tokenization of Descriptions

The vocabulary set or dictionary can be tokenized into individual token numbers or vector representations based on index in set or through the use of predefined sets of vector mappings known as Word Embeddings.

Sequence Generation

Generally, Image Captioning models take in sequential input of words for prediction. The model is inputted the start sequence token followed by the image feature vector. After the first prediction, the output and start sequence combined is given as the next input for the next prediction in sequence.

Image Feature Extraction

For the final step, where the model is inputted with the sequences and image features, we generally use a pretrained image model such as VGG-16, ResNets, etc, to generate the image features. These image features are combined with the sequence vectors and are given as input to the model.

LITERATURE SURVEY

ML and DL based techniques are primarily used to obtain the image characteristics which helps in understanding the image.

ML based approaches will extract handcrafted features from the image using the different techniques like LBP (Ojala et al., 2000), SIFT (Lowe, 2004) and HOG (Dalal & Triggs, 2005). With the help of the features extracted, SVM (Boser, et al., 1992) is used to classify an object. Handcrafted characteristics are task dependent hence cannot be able to extract features from different and diverse variety of data. Real-world data is more complicated with wide variety of images and videos and has a variety of semantic interpretations.

DL based methods will learn features automatically from the data which is used to train them and the importance is that it can easily handle a huge and various range of images and videos. (CNN) (LeCun et al., 1998) is the most commonly used feature learning algorithm, and is followed by Softmax classifier for classification. The RNN is then used to generate the captions with the objects learnt by the CNN.

Neural Image Caption Generator was proposed by Vinyals et al. (Vinyals et al., 2015). (NIC). For image representations, the approach uses a CNN, and for image caption generation, it employs an LSTM. The last layer's output is fed as an input to the LSTM decoder in this special CNN, which uses a novel batch normalization algorithm. This LSTM can maintain track of objects and their previous text descriptions. Maximum likelihood estimation is used to train NIC.

For the challenge of retrieving bidirectional images and words, Karpathy et al. (Karpathy et al., 2014) suggested a deep multimodal approach. The difference between previous approach and this multimodal based approach is that for mapping both image and phrases a common embedding space is used. This approach works at a finer level, embedding image and text fragments. This method argues about the latent, inter-modal alignment of the images by breaking them down into a number of items and words in DTR(De Marneffe et al., 2006). It demonstrates that, when compared to other retrieval methods, the strategy makes considerable improvements in the retrieval job.

For producing innovative image captions, Mao et al. (Mao et al., 2015) suggested a m-RNN technique. A deep RNN for texts and a deep CNN for images are used in this method. The entire m-RNN model is made up of these two sub-networks interacting in a multimodal layer. This function accepts both images and sentence fragments as input. To produce the next word of captions, it estimates the probability distribution. The model contains two word embedding layers, a multimodal, a recurrent and a SoftMax layer .

Kiros et al. (Kiros et al., 2014) suggested a method that uses AlexNet to extract visual information and is based on a Log-Bilinear model. The method of Kiros et al. (Kiros et al., 2014) is closely linked to this m-RNN method. Kiros et al. employ a both fixed length context and temporal context in a recurrent architecture, which allows for context of any length. To build a dense word representation, the two word embedding layers employ a single hot vector. It encodes the words' syntactic as well as semantic meaning.

Another multimodal space-based IC approach was proposed by Chen et al. (Chen & Lawrence Zitnick, 2015). This approach can create image captions by restoring the visual elements from a description. Reference to bidirectional mapping between image and captions is also done. To produce image captions, many existing approaches (Hodosh et al., 2013; Karpathy et al., 2014; Socher et al., 2014) use joint embedding. The reverse projection, which can produce visual characteristics from captions, is not used in the existing approach. This technique dynamically modifies the image's visual representations based on the created words. It also contains an RNN-based visual hidden layer that does reverse projection.

Ren et al. 2017 (Ren et al., 2017) proposed a unique IC approach based on reinforcement learning. This method's architecture consists of two networks that work together to compute the next best word .The "policy network" serves as local direction and aids in predicting the next word based on present conditions. The "value network" serves as a global guide, assessing the reward value while taking into account all conceivable expansions of the existing condition. This approach allows the networks to be adjusted in order to forecast the correct phrases. The caption generated will be very much similar to the ground truth caption.

Another reinforcement learning-based IC system was proposed by Rennie et al. (Rennie et al., 2017). The method prefers test-time inference process for reward normalization instead doing it during training time. It has proved that using test time inference greatly enhances the caption quality.

A fully convolutional localization network design is proposed by Dense captioning (Johnson et al., 2016). In one single forward pass the dense localization layer analyses an image and predicts the region of interest in that image. As a result, unlike Fast R-CNN or Faster R-CNN, it does not require external region proposals.

The localization layer's working idea is based on the work of Faster R-CNN (Ren et al., 2015).

Another dense IC method was proposed by Yang et al. (Yang et al., 2016). These issues can be addressed using this way. First, it deals with an inference technique that is based on both the visual characteristics of the region and the captions projected for that region. The model determines an acceptable bounding box position in the region. Second, they use context fusion to provide a comprehensive semantic description by combining context data with visual aspects of relevant regions.

Mao et al. (Mao et al., 2016) proposed an image-specific text synthesis algorithm. This approach can generate a referencing phrase (FitzGerald et al., 2013; Golland et al., 2010; Kazemzadeh et al., 2014; Mitchell et al., 2010; Mitchell et al., 2013; van Deemter et al., 2006; Viethen & Dale, 2008), which is a description for a certain object or location. It can then deduce the item or location being described using this expression. As a result, the resultant description or expression is fairly clear. A new dataset called ReferIt dataset (Kazemzadeh et al., 2014) is used to address the referring expression.

For image caption creation, Wang et al. (Wang et al., 2016) presented parallel-fusion RNN-LSTM architecture. The method's architecture separates RNN and LSTM hidden units into a number of equal-sized sections. These sections will parallel work with respective ratios for generating captions.

DATASETS

The inputs to the different model architectures for training require large amounts of data. The images are tagged with a set of descriptive caption and made available. These datasets are as follows: Flickr8K, Flickr30K, Visual Genome, MS COCO and Google Conceptual Captions.

Flickr8K

In the Flickr8k dataset, each image has a 5 caption which was obtained through services provided by Amazon Mechanical Turk. Each caption has an accurate description of the objects and the scene depicted in the image, which have an average length of 11.8. For the 8092 images, there are 40,460 caption in the dataset. The usual split for training are 6000 images for training set, 1000 for development set and rest for testing.one sample image along with the captions from the dataset is given below.

1. "A man is doing tricks on a bicycle on ramps in front of a crowd"
2. "A man on a bike executes a jump as part of a competition while the crowd watches"
3. "A man rides a yellow bike over a ramp while others watch"
4. "Bike rider jumping obstacles"
5. "Bike rider jumps off a ramp"

Flickr30K

A total of 158,915 descriptions are included in this dataset, which is an extended version of the Flickr8K. It contains 31783 images with 5 captions per each image. The number of images used for both development and testing is 1000. The rest will be used for training.

Visual Genome

There are more than 100K images in this dataset. Each image consists of 35 objects with dense description annotations and 21 interactions between objects. This dataset is used to train image caption models which involve spatial and semantic relationships.

MSCOCO

Microsoft's research team released this large scale dataset. One of the popular datasets for object detection, instance segmentation and image captioning. A total of about 328,000 images and 2.5 million tags made up this dataset. Each image contains 5 captions.

Google Conceptual Captions

The files contain over 3 millions of images, and are complemented by natural-language caption. The style of the data is similar to the one in the MSCOCO dataset. A dataset of about 3,318,333 train set images, 15,480 validation set images and about 12,559 images as the testing set which is hidden.

EVALUATION METRICS

BLEU

The Bilingual Evaluation Understudy Score(Papineni et al., 2002) is the most used metric of evaluation for image caption tasks. The evaluation score is based on the count of matched n-grams with the reference sentence. The candidate or predicted translation to the n-grams in the reference sentences is the match. A 1-gram is one word whereas a 2-gram or bi-gram is 2 words. The score is calculated to make sure that the translation matches the references.

N-grams are usually continous sets or sequences of words or characters or tokens in a sentences/document. They are basically neighbouring groups of words in a sentence. Based on the number of consecutive occuring words N-grams are classified as unigram (1), bigram (2), trigram (3) and n-gram (3+).

$$\textbf{BLEU} = \min(1, \frac{output - length}{reference - length})(\prod_{i=1}^{4} precision_i)^{\frac{1}{4}}$$

METEOR

The METEOR score(Agarwal & Lavie, 2008) is a metric to evaluate the caption generated with explicit ordering. The score is based on the mean of unigram precision and recall. It includes methods such as stemming and synonym matching. The BLEU score was not good enough and this score makes up for it.

Unigram Precision $\quad P = \dfrac{m}{w_t}$

Unigram Recall $\quad R = \dfrac{m}{w_r}$

Harmonic Mean $\quad F_{mean} = \dfrac{10PR}{R + 9P}$

Penalty $\quad p = 0.5 + (\dfrac{c}{U_m})^3$

Final Meteor Score $\quad M = F_{mean}(1 - p)$

CIDEr

CIDEr is based on consensus. The score is a metric of evaluation. This metric is based on consensus. The weight of the TF-IDF vectors is calculated for each n-gram, and then the similarity between the test sentences and the references are calculated.

$$CIDEr(c_i S_i) = \frac{1}{m} \sum_j \frac{g^n(c_i) \cdot g^n(S_{ij})}{\| g^n(c_i) \| \; \| g^n(S_{ij}) \|}$$

SPICE

The SPICE score is based on the mapping of objects, relationships and features in a graph like structure. Tuples of objects are taken and precision /recall is calculated. The combined score of these scores represents the SPICE Score.

Precision $\quad P(c,S) = \dfrac{|T(G(c)) \otimes T(G(S))|}{|T(G(c))|}$

Recall $\quad R(c,S) = \dfrac{|T(G(c)) \otimes T(G(S))|}{|T(G(S))|}$

SPICE Score $\quad SPICE(c,S) = F_1(c,S) = \dfrac{2 \cdot P(c.S) \cdot R(c.S)}{P(c.S) + R(c.S)}$

ROUGE

ROUGE is a set of metrics for evaluating summarization of sentences and text. It is a combination of 3 major sub scores. Rouge-N for n-gram overlap scores, Rougue-L for longest matching sequence using LCS and the Rouge-S for skip-gram occurrence measure. All these three combined form the ROUGE Evaluation suite.

These are the most frequently used evalution metrics for machine translation tasks.

PROPOSED SYSTEM

The general encoder-decoder architecture is used with a few modifications to assist in the generation of accurate predictions.

- Encoder we use a pretrained ResNet152 model for image feature extraction. We take the features extracted from the model before the FCL and pass them through BatchNormalization layer.
- The features from the encoder are then passed into the LSTM along with the caption sequences. The entire step is called as the Decoder.

ResNet152

ResNets in general solve two general issues in training deep convolutional neural networks. One advantage is that they solve the vanishing gradient problem by using identity mapping. This is ensures that the training of deep CNNs can be done by increasing the number of layers without increasing the training error percentage.

The main idea behind the ResNet layer is that the input is concatenated with the output of each layer and is forwarded to the next layer. This ensures the transfer of image features from one layer to another. This is generally known as "a skip connection".

Decoder

We have already given a brief introduction on LSTM Cell. After we extract the image features we combine the image features along with the tokenized sequences of captions into embeddings. These embeddings will be the input to the LSTM layer. The output of the LSTM is then passed on to a linear layer to predict the output sequence.

Figure 6. An example of a ResNet layer. The input of the layer is finally concatenated with the output.

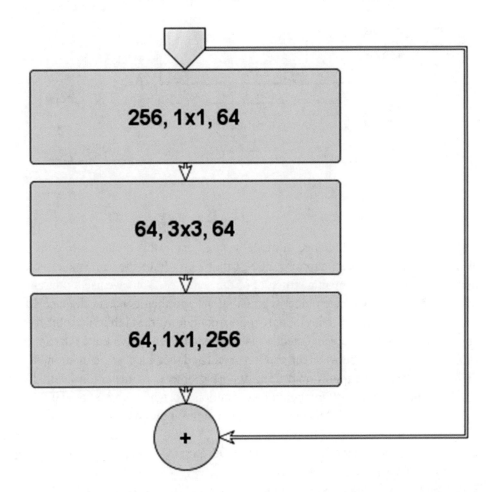

MR-CNN

The Mask RCNN model is deep learning model that solves the problem of combined object detection and instance segmentation. The MR-CNN has a CNN backbone which extracts features from the image. The FPN network generates different size feature maps which is used a region proposal network. Followed by this, the anchor generation module tends to bind different anchors (bounding boxes) to the respective feature maps according to the IntersectionOverUnion (IOU) metric. Another sub model takes these proposed regions and generates the object classes (multi-classified), bounding boxes and masks.

Figure 7. Mask RCNN Architecture Diagram

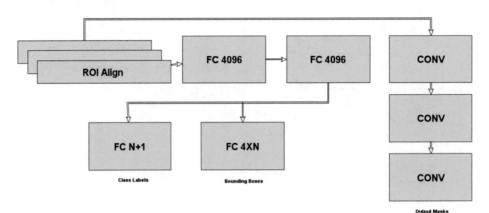

Sometimes during inference, a model trained using this architecture does not always generate the exact captions, atleast they don't have the same exact object match and retain the semantic structure of the sentence. Thus we use a pretrained MR-CNN model for image object segmentations and then generate the labels for each individual object. If the predicted captions containing the highlighted object text and the segmentation / object detection labels match, then we generate the caption. If there is a mismatch between the caption object and the object segmentation label, then we replace the caption object text with the object segmentation label. This similarity implies the working behind an attention mechanism based model.

The replacement of the correct labels is done through the following steps:

- The predicted sentence is exposed to POS Tagging to determine the different parts of speech tag for each word.
- The POS tag for the object segmentation label is also computed.
- The object detection label is then replaced with the wrong caption object by matching the POS tag and then replacing it.

Sometimes multiple types of the same POS tag might be present. For example" a DAT bunch NN of IN different JJ food NN on IN top NN of IN a DAT table NN". Here if we were to replace the word "food" with a particular word for example "pizzas", we will have to first POS tag them as NN and then we will have to replace food with pizzas. But there are multiple words tagged as NN. Thus we generate multiple combination replacements and then use an evaluation function to score the replacements. If the replacement sentence words are not ordered grammatically then they show a low score. The replacement with the highest score is then returned.

The final caption is then sent to the gTTS function to convert the predicted caption as speech for the visually impaired people.

Figure 8. Proposed Image Captioning Model

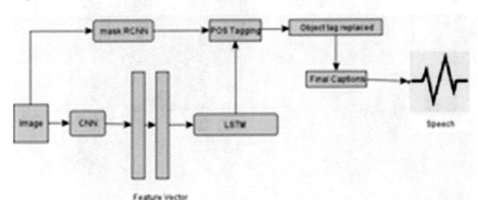

Training Procedure

For the purpose of training, we had leveraged the NVIDIA RTX 2060 GPU on our local desktop. The training process was customized in such way that the best model gets saved for both the Encoder and Decoder models. We had trained the model for 40-50 epochs with 20 steps per epoch. The process took around 5 – 6 hours per training. After training this model, we had taken the images applied the model on them and computed the results.

RESULTS AND DISCUSSION

The base Encoder-Decoder architecture was evaluated for a set of test images. After inference, the different BLEU scores are shown below.

Prediction: a man is holding a cell phone in front of a television
BLEU1: 0.38571428571428573
BLEU2: 0.2463768115942029
BLEU3: 0.14705882352941177
BLEU4: 0.10447761194029852

Figure 9. Ground Truth: A man is holding a phone in front of the atm

Prediction: a person holding a bottle of soda
BLEU1: 0.4782608695652174
BLEU2: 0.29411764705882354
BLEU3: 0.1940298507462687
BLEU4: 0.13636363636363635

Predicted Caption: a bunch of medicines on top of a table
BLEU1: 0.7213114754098361
BLEU2: 0.6
BLEU3: 0.5254237288135594
BLEU4: 0.4482758620689655

Figure 10. Ground Truth: A man is holding a bottle of pepsi

Figure 11. Ground truth: a bunch of different medicines on top of a table

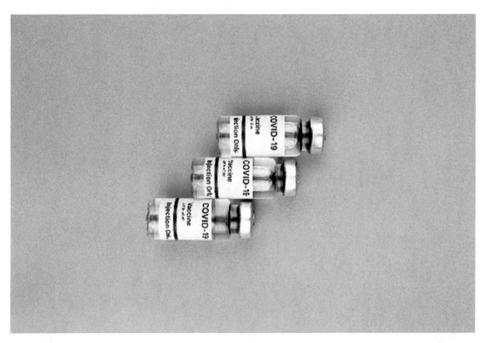

CONCLUSION

An Image captioning method based on the ResNet152 deep learning architecture is implemented in this work. Our proposed method was tested on the Flickr 8K and MSCOCO datasets, and it resulted in much better captioning performance. The caption generated is then converted into speech in any national language. This method has the potential to be implemented into hardware platforms such as smartphones or smart glasses, easing the challenges that visually impaired persons have in their daily lives. Visually impaired people can interpret events in their environment using an NLP incorporated smartphone application with narrator options, making life more enjoyable for them. This could be a promising area for future research.

REFERENCES

Agarwal, A., & Lavie, A. (2008). Meteor, m-bleu and m-ter: Evaluation metrics for highcorrelation with human rankings of machine translation output. In *Proceedings of the Third Workshop on Statistical Machine Translation*. Association for Computational Linguistics. 10.3115/1626394.1626406

Boser,, Guyon, & Vapnik. (1992). A training algorithm for optimal margin classifiers. *Proceedings of the fifth Annual Workshop on Computational Learning Theory*, 144–152.

Chen, X., & Lawrence Zitnick, C. (2015). Mind's eye: A recurrent visual representation for image caption generation. *Proceedings of the IEEE Conference on Computer Vision and Pattern Recognition*, 2422–2431. 10.1109/CVPR.2015.7298856

Dalal, N., & Triggs, B. (2005). Histograms of oriented gradients for human detection. *IEEE Conference on Computer Vision and Pattern Recognition, 1*, 886–893. 10.1109/CVPR.2005.177

De Marneffe, MacCartney, & Manning. (2006). Generating typed dependency parses from phrase structure parses. *Proceedings of LREC, 6*, 449–454.

FitzGerald, N., Artzi, Y., & Zettlemoyer, L. (2013). Learning distributions over logical forms for referring expression generation. *Proceedings of the 2013 Conference on Empirical Methods in Natural Language Processing*, 1914–1925.

Girshick, R. (2015). Fast r-cnn. *Proceedings of the IEEE International Conference on Computer Vision*, 1440–1448.

Golland, D., Liang, P., & Klein, D. (2010). A game-theoretic approach to generating spatial descriptions. In *Proceedings of the 2010 Conference on Empirical Methods in Natural Language Processing*. Association for Computational Linguistics.

Hochreiter, S., & Schmidhuber, J. (1997). Long short-term memory. *Neural Computation, 9*(8), 1735–1780. doi:10.1162/neco.1997.9.8.1735 PMID:9377276

Hodosh, M., Young, P., & Hockenmaier, J. (2013). Framing image description as a ranking task: Data, models and evaluation metrics. *Journal of Artificial Intelligence Research, 47*, 853–899. doi:10.1613/jair.3994

Johnson, J., Karpathy, A., & Li, F.-F. (2016). Densecap: Fully convolutional localization networks for dense captioning. *Proceedings of the IEEE Conference on Computer Vision and Pattern Recognition*, 4565–4574. 10.1109/CVPR.2016.494

Karpathy, A., Joulin, A., & Fei F Li, F. (2014). *Deep fragment embeddings for bidirectional image sentence mapping*. Advances in Neural Information Processing systems.

Kazemzadeh, Ordonez, Matten, & Berg. (2014). ReferItGame: Referring to Objects in Photographs of Natural Scenes. *Proceedings of the Conference on Empirical Methods in Natural Language Processing (EMNLP)*, 787–798.

Kiros, R., Salakhutdinov, R., & Zemel, R. (2014). Multimodal neural language models. *Proceedings of the 31st International Conference on Machine Learning (ICML-14)*, 595–603.

LeCun, Y., Bottou, L., Bengio, Y., & Haffner, P. (1998). Gradient-based learning applied to document recognition". *Proceedings of the IEEE, 86*(11), 2278–2324. doi:10.1109/5.726791

Lowe, D. (2004). Distinctive image features from scale-invariant keypoints. *International Journal of Computer Vision, 60*(2), 91–110. doi:10.1023/B:VISI.0000029664.99615.94

Mao, J., Huang, J., Toshev, A., & Camburu, O. (2016). Generation and comprehension of unambiguous object descriptions. *Proceedings of the IEEE Conference on Computer Vision and Pattern Recognition*, 11–20.

Mao, J., Xu, W., Yang, Y., Wang, J., Huang, Z., & Yuille, A. (2015). Deep captioning with multimodal recurrent neural networks (m-rnn). *International Conference on Learning Representations (ICLR)*.

Mitchell, M., van Deemter, K., & Reiter, E. (2010). Natural reference to objects in a visual domain. In *Proceedings of the 6th International Natural Language Generation Conference*. Association for Computational Linguistics.

Mitchell, M., Van Deemter, K., & Reiter, E. (2013). *Generating Expressions that Refer to Visible Objects*. HLT-NAACL.

Ojala, T., Pietikäinen, M., & Mäenpää, T. (2000). Gray scale and rotation invariant texture classification with local binary patterns. In *European Conference on Computer Vision*. Springer. 10.1007/3-540-45054-8_27

Papineni, K., Roukos, S., Ward, T., & Zhu, W.-J. (2002). BLEU: A method for automatic evaluation of machine translation. In *Proceedings of the 40th Annual Meeting on Association for Computational Linguistics*. Association for Computational Linguistics.

Ren, S., He, K., Girshick, R., & Sun, J. (2015). *Faster R-CNN: Towards real-time object detection with region proposal networks*. Advances in Neural Information Processing Systems.

Ren, Z., Wang, X., Zhang, N., Lv, X., & Li, L.-J. (2017). Deep Reinforcement Learningbased Image Captioning with Embedding Reward. *Proceedings of the IEEE Conference on Computer Vision and Pattern Recognition (CVPR)*, 1151–1159.

Rennie, Marcheret, Mroueh, Ross, & Goel. (2017). Self-critical sequence training for image captioning. *Proceedings of the IEEE Conference on Computer Vision and Pattern Recognition (CVPR)*, 1179–1195.

Socher, R., Karpathy, A., & Quoc, V. (2014). Grounded compositional semantics for finding and describing images with sentences. *Transactions of the Association for Computational Linguistics*, 2, 207–218. doi:10.1162/tacl_a_00177

van Deemter, K., van der Sluis, I., & Gatt, A. (2006). Building a semantically transparent corpus for the generation of referring expressions. In *Proceedings of the Fourth International Natural Language Generation Conference*. Association for Computational Linguistics. 10.3115/1706269.1706296

Viethen, J., & Dale, R. (2008). The use of spatial relations in referring expression generation. In *Proceedings of the Fifth International Natural Language Generation Conference*. Association for Computational Linguistics. 10.3115/1708322.1708334

Vinyals, O., Toshev, A., Bengio, S., & Erhan, D. (2015). Show and tell: A neural image caption generator. *Proceedings of the IEEE Conference on Computer Vision and Pattern Recognition*, 3156–3164. 10.1109/CVPR.2015.7298935

Wang, M., Song, L., Yang, X., & Luo, C. (2016). A parallel-fusion RNN-LSTM architecture for image caption generation. In *2016 IEEE International Conference on Image Processing (ICIP)*. IEEE. 10.1109/ICIP.2016.7533201

Yang, L., Tang, K., Yang, J., & Li, L.-J. (2016). Dense Captioning with Joint Inference and Visual Context. *Proceedings of the IEEE Conference on Computer Vision and Pattern Recognition (CVPR)*, 1978–1987.

Chapter 12
Statistical Hypothesization and Predictive Modeling of Reactions to COVID-19- Induced Remote Work:
Study to Understand the General Trends of Response to Pursuing Academic and Professional Commitments

Arjun Sharma
Vellore Institute of Technology, Chennai, India

Om Prakash Swain
Vellore Institute of Technology, Chennai, India

Hemanth Harikrishnan
Vellore Institute of Technology, Chennai, India

Utkarsh Utkarsh
Vellore Institute of Technology, Chennai, India

Sathiya Narayanan Sekar
Vellore Institute of Technology, Chennai, India

Akshay Giridhar
Vellore Institute of Technology, Chennai, India

ABSTRACT

The initial outbreak of the coronavirus was met with lockdowns being enforced all over the world in March 2020. A prominent change in human lifestyle is the shift of professional and academic work to online platforms, as opposed to previously attending to them in person. As with any major change, the implementation of complete remote work and study is expected to affect different people differently. Through the results of a questionnaire designed as per the implications of the self-

DOI: 10.4018/978-1-6684-3843-5.ch012

Copyright © 2023, IGI Global. Copying or distributing in print or electronic forms without written permission of IGI Global is prohibited.

efficacy theory shared with people who were either students, working professionals, entrepreneurs, or homemakers aged between 12 and 60 years, the authors perform statistical analysis and subsequently hypothesize how different aspects of remote work affect the population from a mental standpoint using t-test, with respect to their professional or academic work. This is followed by predictive modelling through machine learning algorithms to classify working preference as 'remote' or 'in-person'.

INTRODUCTION

COVID-19 has spread across the world, with the World Health Organization terming it a 'pandemic' in March 2020 (World Health Organization, 2020). It has infected over 238 million across the globe, killing at least 4.8 million as of October 2021 (Worldometers.info. 2021). To reduce the spread and impact of the disease, government sanctioned lockdowns were imposed. The lockdowns were accompanied by the adoption of remote work and study alternatives.

This work focuses on understanding the effect of various aspects of complete remote work through means of a survey answered by 450 respondents aged between 12 and 60 years across India. At the time of data collection, all the participants were WFH full-time. Participation in the research was voluntary, anonymous, and without any reward. The process of data collection was in full compliance of the declaration of Helsinki (World Medical Association Declaration of Helsinki, 2013). The questionnaire comprises of questions modelled on the basis of the implications of the self-efficacy theory and its 4 main constructs described in section II. The data obtained is analyzed to understand factors that influence an individual's ability to work, how their state of mind was affected by new norms, and also their general experience in a virtual setting. Using t-test, the proposed hypotheses are validated for different behavioral aspects showcased by respondents. Finally, predictive modelling using 6 supervised learning models is performed to predict an individual's preferred method of working as either 'remote', or 'in-person'. In this work, section II discusses the Self-Efficacy Theory and other relevant research undertakings. Section III discusses the methods involved for questionnaire preparation. Section IV comprises the exploratory data analysis of the survey results. Section V constitutes the hypotheses proposed by the authors regarding the behavioral experiences of those working remote. Section VI summarizes the results obtained in predictive modelling stage. Section VII concludes the work.

The contributions through this work are, firstly, formulating and testing hypotheses the results of which, can be used to potentially increase the quality of work and level of satisfaction of employees working remotely. Secondly, the usage of predictive analytics in determining the work-pattern (remote or non-remote) preferences of individuals from the set of features discussed in section V.

BACKGROUND

Self-efficacy theory (Bandura, 1986) claims that the behavior, environment, and cognitive factors of an individual share high levels of interrelation. Self-efficacy is defined as "a judgement of one's ability to execute a particular behavior pattern." Further assessment shows that self-efficacy is crucial in levels of motivation and performance exhibited by an individual (Wood & Bandura, 1989), implying that self-efficacy levels also influence the amount of effort and/or time invested on a particular task. Those with higher beliefs of self-efficacy take more effort to complete their tasks, conversely, those with lower self-efficacy beliefs tend to undertake relatively lesser effort, spend less time and sometimes even abandoning it.

According to self-efficacy theory, 4 major sources are considered by an individual through the formation of their self-efficacy judgements, they are: Performance Accomplishment, Vicarious Experience, Social Persuasion and, Physiological and Emotional State (Wrycza & Maslankowski, 2020).

Self-efficacy theory excels in behavior and performance prediction (Bandura, 1978). The theory and evidence supporting its empirical implications are robust and its implications strongly suit the study of virtual organizations (Staples et al., 1999). Employees working remote have little support or guidance, strongly fitting the current situation across the world due to COVID-19.

In addition, the authors have compared their work with that of other research groups. In (Zhang C et al., 2021; Galanti T et al., 2021), the general opinion of the public regarding remote-work due to COVID-19 enforced lockdowns was gathered through the online microblogging website "Twitter." Over 500,000 'tweets' (Microblogs of less than 140 characters) were analyzed. They perform time-series analysis to plot the frequency of remote-work related tweets over time. Sentiment analysis is performed to understand to what extent remote-work is embraced. Over 50% of the tweets are in favor of remote work, about 40% tweets have a neutral stance, and about 7.5% tweets are negative. While the study highlights the general proportion of sentiment in favor or against remote work, it does not highlight such intricacies across other demographics such as age or gender. It also does not account for different work patterns and influence of remote-work on the mindset of people.

The authors in (World Health Organization, 2021) model the intricacies of daily life while working from home by exploring the influence of family-work conflict, social isolation, distracting environment, job autonomy, and self-leadership have on employees' productivity, work engagement, and stress. However, this study does not include anyone who is not an employee such as students, entrepreneurs, etc. In addition, the study also does not involve a means to predict one's preference of working in a particular setting, over others.

Overall, the existing literature either does not highlight the experience of individual other than employees, in addition, they not involve a predictive modelling approach towards determining factors that could or could not contribute to the preferences of working from home or working in-person which are highlighted in this work.

METHOD FOR QUESTIONNAIRE FRAMING

The appendix of this work has the questions that were part of the questionnaire that was shared with participants. While the self-efficacy theory broadly covers possible questions that participants may be asked, it was important to design questions based on existing evidence for potential of altering one's approach to working remote. Age, gender, and occupation (i, ii, iii in appendix respectively), were asked to establish trends across demographics during the exploratory data analysis phase. In addition, the considerations given by WHO (Chung et al., 2020), were also taken into consideration for question formation.

Research undertaken by other research groups shows that professionals tend to spend more hours working while working from home (Gibbs et al., 2021), to verify the occurrence of the same, questions iv and v (in appendix) were asked. Time spent travelling was considered part of time spent working for the reason that this travel on a daily basis was to pursue their professional commitments in their office/institution.

In regards to productivity levels, other research groups (World Health Organization, 2020; Zhang C et al, 2021) report a decrease in general productivity among the working population. Productivity being an important metric for measuring self-efficacy, makes question vii important in WFH context.

As specified by (Chung et al., 2020; Gibbs et al, 2021), sleep cycles and patterns were generally observed to be disturbed. Question viii was imperative to understand the sleep patterns changes (if any), in the general opinion of the answerers of the questionnaire.

With the lockdown, the concerns on mental health, picking up hobbies, in addition to family interactions, and taking care of one's own well-being would be important considerations (Yuksel D et al., 2021; Conroy DA et al., 2021; Morse KF et al., 2021). These evidences make it necessary to ask questions ix, x, xi, xii and xvi (in appendix).

The complete migration to virtual platforms requires getting acclimatized to the online platforms and media of communication online. Considering this and those in (Morelli M et al., 2021), it necessitates asking the question xiii (in appendix).

Another important consideration is whether the home environment allows one to focus on their commitments virtually, (Rehman et al., 2021) mentions the possibility of distractions at home not allowing one to function with appreciable focus, making question xiv important from WFH standpoint.

To understand the overall preference in work setting (either remote or in-person), question xxii (in appendix), was asked. This would also serve as the target variable for the predictive analytics procedure as described in section VI.

EXPLORATORY DATA ANALYSIS

1. Daily Time Commitment

Table I represents age wise trends on the per-day average time spent on work. The age group between 12 to 18 years old shows a decrease of 1.7 hours of time commitment per day. This can be attributed to zero travel time and less time preparing for school. A similar trend is observed for those between ages 19 and 25, as seen in Table 1. Another reason for a sharp decrease in time spent working on average could be the result of widespread loss of employment, especially for those between ages 18 and 24 (Gould et al., 2020).

Respondents between 26 and 32 years on average, spent more time working at home than they spent travelling and working combined on a regular day. It is observed that except for students, everyone else spent more time working at home. Notably, homemakers saw the largest increase in their working hours during the pandemic.

Table 1. Daily time commitment across ages before and during the pandemic

Age Group	Pre-Pandemic Work Hours (Work + Travel)	Work Hours at Home
12-18	7.4	5.7
19-25	7.9	6.8
26-32	8.7	9.4
33-40	8.3	7.4
41-50	9.6	8.7
51-60	8	9.45

2. Adjusting to New Setting

Respondents rated their ease in adjusting to an online environment from 1 to 5 in increasing order of difficulty. From Table 2, males found it tougher migrating to online environment than females except for those in aged between 19-25 and 26-32. Those between 26-32 years found it easiest to migrate to online platforms for both genders. Said age group corresponds to a relatively early stage of one's career, implying that those in early career stages are more adept at working online, while those aged between 33-40 years found it most difficult to make the same adjustment for both genders. From Table 3, homemakers found it the easiest to adapt to an online work environment, followed by entrepreneurs. Students in general experience difficulty.

Table 2. Trends across gender and age pertaining to the ease in adjusting to an online setting

Age	Gender	Ease of Adjustment
12-18	Male	3.41
	Female	2.64
19-25	Male	2.48
	Female	2.79
26-32	Male	1.91
	Female	1.97
33-40	Male	3.56
	Female	3.94
41-50	Male	2.32
	Female	1.87
51-60	Male	2.80
	Female	2.66

Table 3. Trends across occupation pertaining to the ease in adjusting to an online setting

Occupation	Ease of Adjustment
Student in School	2.61
Student in College	2.73
Entrepreneur	2.08
Homemaker	1.49
Working Professional	2.43

3. Effect of Home Environment

Respondents rated the level of disturbance they encountered during work hours from 1 to 5 in increasing order of noisiness, where 1 signified a quiet environment, and 5 signified a highly chaotic and noisy environment. From Figure 1, the level of disturbance experienced by respondents follows a normal distribution with mean 2.91 (implying a mostly neutral outlook towards level of noise at home), and standard deviation 1.198.

Figure 1. Disturbance level distribution experienced by respondents working remote

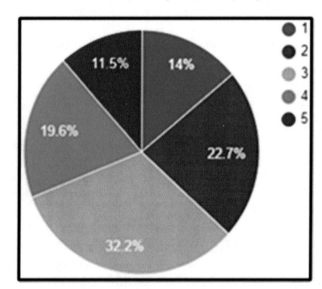

4. Changes in Productivity

Respondents rated their level of productivity while working remote from 1 to 5 In increasing order. 1 signified no productivity change, and 5 signified substantial increase. From Figure 2, there is a slight disagreement to this statement. The responses follow a normal distribution with mean 2.81, and standard deviation of 1.132.

Figure 2. Distribution of agreement level for increase in productivity

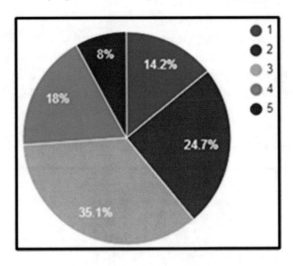

5. Changes in Sleep Cycle

Respondents rated their level of agreement to whether their sleep cycle has improved working remote from 1 to 5 in increasing order of agreement. As per Figure 3, there is disagreement. The responses follow a normal distribution with mean 2.54 and standard deviation 1.218.

Figure 3. Distribution of agreement level for change in sleep cycle

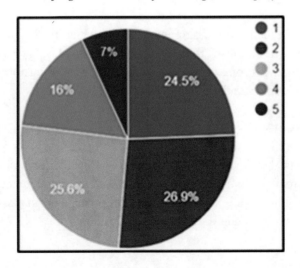

6. New Hobbies for Recreation

Respondents rated their level of agreement with whether they were able to pick up new hobbies while working remote from 1 to 5 in increasing order of agreement. As per Figure 4, there is agreement to this statement. The response follows a normal distribution, with a mean of 3.2, and standard a deviation of 1.289.

Figure 4. Distribution of agreement level for picking up new hobbies

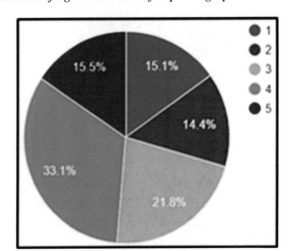

7. Time Spent with Family

Respondents rated their level of agreement with whether they were spending more time with family while working remote from 1 to 5 in increasing order of agreement. From Figure 5, there is strong agreement. The response follows a normal distribution, with mean 3.51, and a standard deviation of 1.293.

8. Stress Levels

Respondents rated their level of agreement with whether they experience lower stress while working remote from 1 to 5 in increasing order of agreement. As per Figure 6, the response indicates slight disagreement, following a normal distribution, with a mean of 2.85, and a standard deviation of 1.256.

Figure 5. Distribution of agreement level for time spent with family

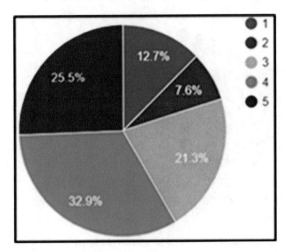

Figure 6. Distribution of agreement level for Stress level reduction

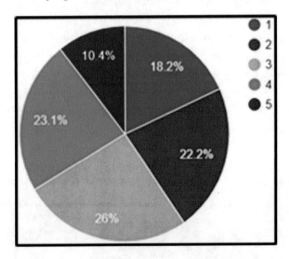

9. Time Spent on Self-Care

Respondents rated their level of agreement with whether they were spending time on their self-care while working remote from 1 to 5 in increasing order of agreement. As per Figure 7, there is agreement. The response follows a normal distribution, with a mean of 3.22, and a standard deviation of 1.445.

Figure 7. Distribution of agreement level for spending time on self-care

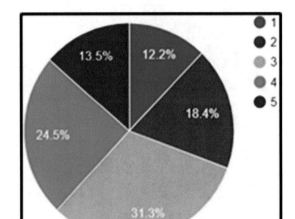

10. Overall Working Pattern Preference

Respondents were asked to choose their overall working pattern preference from either working/studying remotely or working/studying in-person (physically attending to their work or their classes). From Figure 8 it is evident that the majority of the respondents prefer to go to their workplace or institution and attend to their commitments in-person, rather than pursuing it online, with 72.2% respondents preferring in-person work/classes, and 27.8% preferring to do so virtually.

Figure 8. Distribution of overall working preference (remote or in-person)

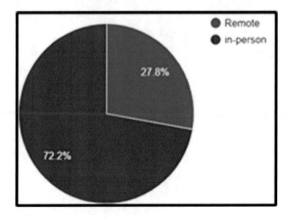

Table 4 highlights correlation observed between constructs that influence self-efficacy in a remote-work setting from the data obtained from the answers given by respondents to the questionnaire. Table 4 presents 8 metrics and their mutual level of correlation. Based on the correlation levels for any 2 metrics, those with significant absolute values are chosen. 2 tailed paired t-test is performed for validation of Hypothesis at p=0.05.

Table 4. Table representing correlation between different behavioral constructs while pursuing occupational commitments virtually

	1.	**2.**	**3.**	**4.**	**5.**	**6.**	**7.**	**8.**
1. Difficulty in Adjusting to Online Work	1	0.51	-0.64	-0.054	-0.24	-0.14	-0.26	-0.21
2. Noise in Home Environment	0.51	1	-0.37	-0.24	-0.18	-0.18	-0.26	-0.19
3. Productivity Working Remote	-0.64	-0.37	1	0.14	0.31	0.21	0.29	0.23
4. Balanced Sleep Hours	-0.054	-0.24	0.14	1	0.56	0.56	0.63	0.55
5. Picking up Hobbies	-0.24	-0.18	0.31	0.56	1	0.71	0.69	0.69
6. Time Spent with Family	-0.14	-0.18	0.21	0.56	0.71	1	0.68	0.71
7. Tendency to lower stress	-0.26	-0.26	0.29	0.63	0.69	0.68	1	0.72
8. Time spent on self-care	-0.21	-0.19	0.23	0.55	0.69	0.71	0.72	1

Hypothesis One: Increased Difficulty in Adjusting to Online Environment is associated with Noisy Home Environments

From Table 4, the difficulty in adjusting to an online environment seems to share a linear relationship with the level of noise at home, with a correlation coefficient of **0.51**. Consider:

H_0: There is no significant evidence for a linear relationship between increased difficulty in adjusting to an online environment and noisy home environment.

H_A: There is significant evidence for a linear relationship between increased difficulty in adjusting to an online environment and noisy home environment.

Hypothesis test outcome:

Table 5. Hypothesis test outcome for Hypothesis One

t	df	p	95% CI
1.4747	449	0.1412	-0.03 to – 0.22

Since p>0.05, H_A is rejected and H_0 is Accepted. There is no significant evidence for a linear relationship between noisy home environment and difficulty in adjusting to an online environment.

Hypothesis Two: Picking up hobbies is associated with balanced sleep when working remote

From Table 4, building new hobbies seems to share a linear relationship with balanced sleep, with a correlation coefficient of **0.56**. Consider:

H_0: There is no significant evidence for a linear relationship between picking up hobbies and balanced sleep.

H_A: There is significant evidence for a linear relationship between picking up hobbies and balanced sleep.

Hypothesis test outcome:

Table 6. Hypothesis test outcome for Hypothesis Two

t	df	p	95% CI
9.3672	449	3.6354e^{-19}	0.51 to 0.79

Since p<0.05, H_0 is rejected and H_A is Accepted. There is significant evidence for a linear relationship between picking up hobbies and balanced sleep.

Hypothesis Three: Time Spent with family is associated with picking up hobbies when working remote

From Table 4, time spent with family seems to share a linear relationship with picking up hobbies, with a correlation coefficient of **0.71**. Consider:

H_0: There is no significant evidence for a linear relationship between time spent with family and picking up hobbies.

H_A: There is significant evidence for a linear relationship between time spent with family and picking up hobbies.

Hypothesis test outcome:

Table 7. Hypothesis test outcome for Hypothesis Three

t	df	p	95% CI
5.7946	449	$1.2907e^{-8}$	0.21 to 0.42

Since p<0.05, H0 is rejected and HA is Accepted. There is significant evidence for a linear relationship between time spent with family and picking up hobbies.

Hypothesis Four: Time spent with family is associated with Balanced Sleep when working remote

From Table 4, time spent with family seems to share a linear relationship with balanced sleep cycles, with a correlation coefficient of **0.56**. Consider:

H_0: There is no significant evidence for a linear relationship between time spent with family and having a balanced sleep.

H_A: There is significant evidence for a linear relationship between time spent with family and having a balanced sleep.

Hypothesis test outcome:

Table 8. Hypothesis test outcome for Hypothesis Four

t	df	p	95% CI
9.5543	449	$1.7868e^{-39}$	0.672 to 1.012

Since p<0.05, H0 is rejected and HA is Accepted. There is significant evidence for a linear relationship between time spent with family and having a balanced sleep.

Hypothesis Five: Time spent on self-care is associated with lower stress when working remote

From Table 4, time spent on self-care seems to share a linear relationship with lower stress levels, with a correlation coefficient of **0.72**. Consider:

H$_0$: There is no significant evidence for a linear relationship between time spent on self-care and lower stress levels.

H$_A$: There is significant evidence for a linear relationship between time spent on self-care and lower stress levels.

Hypothesis test outcome:

Table 9. Hypothesis test outcome for Hypothesis Five

t	df	p	95% CI
5.0201	449	7.4597e^{-7}	0.14 to 0.32

Since p<0.05, H0 is rejected and HA is Accepted. There is significant evidence for a linear relationship between time spent on self-care and lower stress levels.

Hypothesis Six: Low Levels of Stress are related with a Balanced Sleep when working remote

From Table 4, lower stress seems to share a linear relationship with a balanced sleep, with a correlation coefficient of **0.63**. Consider:

H$_0$: There is no significant evidence for a linear relationship between lower stress levels and balanced sleep.

H$_A$: There is significant evidence for a linear relationship between lower stress levels and balanced sleep.

Hypothesis test outcome:

Table 10. Hypothesis test outcome for Hypothesis Six

t	df	p	95% CI
4.9272	449	2.377e^{-16}	0.19 to 0.43

Since p<0.05, H0 is rejected and HA is Accepted. There is significant evidence for a linear relationship between lower stress levels and balanced sleep.

Hypothesis Seven: Picking up Hobbies is Associated with Lower Stress when working remote

From Table 4, picking up hobbies while working remote sems to share a linear relationship with lower stress levels, with a correlation coefficient of **0.69**. Consider:

H_0: There is no significant evidence for a linear relationship between picking up hobbies and lower stress levels.

H_A: There is significant evidence for a linear relationship between picking up hobbies and lower stress levels.

Hypothesis test outcome:

Table 11. Hypothesis test outcome for Hypothesis Seven

t	df	p	95% CI
3.5612	449	$1.139e^{-8}$	0.20 to 0.69

Since $p<0.05$, H0 is rejected and HA is Accepted. There is significant evidence for a linear relationship between picking up hobbies and lower stress levels.

Hypothesis Eight: Time spent on self-care is associated with picking up Hobbies when working remote

Table 4 suggests that time spent on self-care shares a linear relationship with picking up new hobbies, with a correlation coefficient of **0.69**. Consider:

H_0: There is no significant evidence for a linear relationship between time spent on self-care and picking up hobbies.

H_A: There is significant evidence for a linear relationship between time spent on self-care and picking up hobbies.

Hypothesis test outcome:

Table 12. Hypothesis test outcome for Hypothesis Eight

t	df	p	95% CI
2.2442	449	0.0253	0.01to 0.20

Since $p<0.05$, H0 is rejected and HA is Accepted. There is significant evidence for a linear relationship between time spent on self-care and picking up hobbies.

MACHINE LEARNING BASED PREDICTIVE MODELLING

1. Data Preprocessing

Data obtained from respondents is encoded through the Scikit-learn library in Python (Pedregosa et al., 2011). The categorical features were One-Hot encoded; different categorical values were converted to numerical values, further normalized to the range from -1 to 1, minimizing influence of outliers in the predictions. With respect to dependent features Label encoder was used as the values are cardinal. To train the model, the data was split into train and test set. To reduce the effect outliers Standard scaling is performed. Next, Principal component analysis is used to reduce the dimensions of independent features. The dataset does not contain duplicate or missing values because all questions were compulsory to answer, and respondents could answer the survey only once.

2. Work Setting Preference Model

The prediction of preference of remote work or working in-person is a binary classification problem, and hence requires the use of supervised learning models. The interest in the implementation of these models is to detect as to whether or not the individual would prefer working/studying remote over studying/working in-person. The following classifier models were used: Logistic Regression, K Nearest Neighbor, Random Forest, Naïve Bayes, XGBoost, Passive Aggressive Classifier, and Support Vector Classifier, also using the scikit-learn library. The dataset was split into a ratio of 80:20 corresponding to training and testing sets respectively.

3. Machine Learning Based Predictive Models

i.   **Logistic Regression**: A linear model for binary classification using the logistic function to model the dependent variables (King & Zeng, 2001).
ii.   **Random Forest:** A tree-based bagging algorithm where the successive trees are created by usage of different samples in the dataset. For prediction purposes, the simple majority vote is taken (Liaw & Wiener, 2002).
iii. **Naïve Bayes:** It is a supervised machine learning model used for classification tasks. It is a probabilistic model based on Bayes Theorem (Webb G.I., 2011).

iv. **XGBoost:** A tree-based gradient boosting algorithm uses a set of weak learners. After fitting these weak learners, the final prediction is produced by the combination of the predictions of the weak learners through weighted sum (Chen & Guestrin, 2016).

v. **Passive Aggressive Classifier:** This is an online learning model, making it very effective in situations with continuous data input. It remains passive in case of a correct prediction, and responds aggressively to incorrect predictions (Crammer et al., 2006).

vi. **Support Vector Classifier**: An algorithm used for classification that may or may not be linear. The algorithm uses hyperplanes in higher dimensional space to perform class separation (Smola & Schölkopf, 2004).

4. Experimentation Results

The implemented models are compared to ascertain the most appropriate for classifying between virtual and in-person work preference. The Model uses 18 features: Age, Gender, Occupation, Daily Work Hours Before Pandemic, Daily Work Hours During Pandemic, Average Daily Travel Time, Difficulty in Adjusting to Online Environment, Home Noise Levels, Productivity at Home, Balanced Sleep Levels, New Hobbies, Time Spent with Family, Stress Level, Time Spent on Self-Care, Appealing Factors of Remote Work, Negatives of Remote Work, and overall work preference. As in Table VI, XGBoost Classifier model performs best for the dataset gathered through the survey in terms of accuracy and the F1 score.

Table 13. Performance evaluation of the models using the real dataset

Classifier Model	Accuracy (%)	F1 Score
Logistic Regression	92.34	0.92
Random Forest	83.83	0.86
Naïve Bayes	88.09	0.88
XGBoost	**94.04**	**0.94**
Passive Aggressive	91.49	0.92
Support Vector	93.19	0.93

CONCLUSION

This study ascertains different influences on one's mindset in a remote setting using statistical analysis techniques and predictive modelling. Contributions can be summarized in three points: 1) Proposal of hypotheses that relate different aspect of remote work and how they influence an individual's mindset, 2) Predictive modelling to ascertain the tendency of preferring one form of work over another, and 3) The usage of the results of the above hypotheses to alter to an extent, the approach in such spheres to create a productive atmosphere.

FUTURE RESEARCH DIRECTIONS

This study was aimed at the general population to summarize the response towards having to pursue academic and professional commitments from a remote setting. While the authors believe that this work is effective in its objective, there is tremendous scope to perform a similar study by focusing on specific groups of interests in this scenario such as medical professionals and other essential workers whose lifestyle is vastly different from others in the current scenario surrounding the pandemic.

ACKNOWLEDGMENT

The authors would like to express their deepest gratitude to the respondents of the questionnaire, without the data contributed by them, this study and its implications would not have been possible. This research received no specific grant from any funding agency in the public, commercial, or not-for-profit sectors.

REFERENCES

Bandura, A (1986), The Explanatory and Predictive Scope of Self-Efficacy Theory. *Journal of Social and Clinical Psychology, 4.*

Bandura, A. (1978). Reflections on self-efficacy. *Advances in Behaviour Research and Therapy, 1*(4), 237-269.

Conroy, D. A., Hadler, N. L., & Cho, E. (2021). The effects of COVID-19 stay-at-home order on sleep, health, and working patterns: A survey study of US health workers. *Journal of Clinical Sleep Medicine, 17*(2), 185–191.

De', R., Pandey, N., & Pal, A. (2020). Impact of digital surge during Covid-19 pandemic: A viewpoint on research and practice. *International Journal of Information Management, 55*, 102171. doi:10.1016/j.ijinfomgt.2020.102171

Galanti, T., Guidetti, G., Mazzei, E., Zappalà, S., & Toscano, F. (2021). Work From Home During the COVID-19 Outbreak: The Impact on Employees' Remote Work Productivity, Engagement, and Stress. *Journal of Occupational and Environmental Medicine, 63*(7), e426–e432. doi:10.1097/JOM.0000000000002236

Gibbs, M., Mengel, F., & Siemroth, C. (2021). *Work from Home & Productivity: Evidence from Personnel & Analytics Data on IT Professionals.* University of Chicago, Becker Friedman Institute for Economics Working Work No. 2021-56. Available at SSRN: https://ssrn.com/abstract=3843197

Gould, E., & Kassa, M. (2020). *Young workers hit hard by the COVID-19 economy.* Economic Policy Institute. https://www.epi.org/publication/young-workers-covid-recession/

King, G., & Zeng, L. (2001). Logistic regression in rare events data. *Political Analysis, 9*(2), 137–163.

Liaw, A., & Wiener, M. (2002, December). Classification and regression by randomForest. *R News, 2*, 18–22.

Morelli, M., Cattelino, E., Baiocco, R., Trumello, C., Babore, A., Candelori, C., & Chirumbolo, A. (2020). Parents and Children During the COVID-19 Lockdown: The Influence of Parenting Distress and Parenting Self-Efficacy on Children's Emotional Well-Being. *Frontiers in Psychology, 11*, 584645. doi:10.3389/fpsyg.2020.584645

Morse, K. F., Fine, P. A., & Friedlander, K. J. (2021). Creativity and Leisure During COVID-19: Examining the Relationship Between Leisure Activities, Motivations, and Psychological Well-Being. *Frontiers in Psychology, 12*, 609967. doi:10.3389/fpsyg.2021.609967

Peters, P., Tijdens, K. G., & Wetzels, C. (2004). 'Employees' opportunities, preferences, and practices in telecommuting adoption'. *Information & Management, 41*(4), 469–482.

Rehman, U., Shahnawaz, M. G., & Khan, N. H. (2021). Depression, Anxiety and Stress Among Indians in Times of Covid-19 Lockdown. *Community Mental Health Journal, 57*(1), 42–48. doi:10.100710597-020-00664-x

Smola, A. J., & Schölkopf, B. (2004). A tutorial on support vector regression. *Statistics and Computing*, *14*, 199–222. https://doi.org/10.1023/B:STCO.0000035301.49549.88

Staples, D. S., Hulland, J. S., & Higgins, C. A. (1999). A Self-Efficacy Theory Explanation for the Management of Remote Workers in Virtual Organizations. *Organization Science*, *10*(6), 758–776. https://dx.doi.org/10.1287/orsc.10.6.758

Webb, G. I. (2011). Naïve Bayes. In C. Sammut & G. I. Webb (Eds.), Encyclopedia of Machine Learning. Springer. https://doi.org/10.1007/978-0-387-30164-8_576.

WMA Declaration of Helsinki. (n.d.). https://www.wma.net/policies-post/wma-declaration-of-helsinki-ethical-principles-for-medical-research-involving-human-subjects/

Wood, R., & Bandura, A. (1989). Social Cognitive Theory of Organizational Management. *The Academy of Management Review, 14*(3), 361–384. www.jstor.org/stable/258173

Working from home during the COVID-19 lockdown: Changing preferences and the future of work. (n.d.). https://www.birmingham.ac.uk/Documents/college-social-sciences/business/research/wirc/epp-working-from-home-COVID-19-lockdown.pdf

World Health Organisation. (2020, March 11). *Coronavirus disease 2019 (COVID-19) Situation Report – 51*. https://www.who.int/docs/default-source/coronaviruse/situation-reports/20200311-sitrep-51-covid-19.pdf?sfvrsn=1ba62e57_10

World Health Organization. (2020a). *Mental Health and Psychosocial Considerations During the COVID-19 Outbreak*. Available at: https://www.who.int/docs/default-source/coronaviruse/mental-health-considerations.pdf?sfvrsn=6d3578af_2

Wrycza, S., & Maślankowski, J. (2020). Social Media Users' Opinions on Remote Work during the COVID-19 Pandemic. Thematic and Sentiment Analysis. *Information Systems Management*, *37*(4), 288–297. doi:10.1080/10580530.2020.1820631

Yuksel, D., McKee, G. B., & Perrin, P. B. (2021). Sleeping when the world locks down: Correlates of sleep health during the COVID-19 pandemic across 59 countries. *Sleep Health*, *7*(2), 134–142. doi:10.1016/j.sleh.2020.12.008

Zhang, C., Yu, M. C., & Marin, S. (2021, June). Exploring public sentiment on enforced remote work during COVID-19. *The Journal of Applied Psychology*, *106*(6), 797–810. doi:10.1037/apl0000933

KEY TERMS AND DEFINITIONS

Accuracy: The number of correct prediction divided by the total number of predictions.

CI: Confidence Interval, it measures the degree of certainty in a sampling method. The most common probability limit is 95%, which is also the limit specified in this work.

df: The largest number of logically independent values, that can vary within the dataset.

F1 Score: Harmonic mean of the recall and precision.

H0: Null Hypothesis, the true difference between the group means is zero.

HA: Alternate Hypothesis, the true difference between the group means is not zero.

p: Probability that the sample data results occurred by chance.

Precision: Number of true positives divided by sum of the number of true positives and the number of false positives.

Recall: Number of true positives divided by the sum of the number of true positives and number of false negatives.

t: Calculated difference in units of standard error.

APPENDIX

Questionnaire Items

i. Age (12-18, 19-25, 25-32, 33-40, 41-50, 51-60)
ii. Gender (Male, Female, Other)
iii. Occupation (Working Professional, Student in School, Student in College, Homemaker, Entrepreneur)
iv. Daily average work hours before remote work enforcement
v. Daily average work hours after remote work enforcement
vi. Daily average time spent travelling

(Questions **vii.** To **xv.** are answered with the options: Strongly Disagree, Disagree, Neutral, Agree, Strongly Agree)

vii. My Productivity levels have increased working remote
viii. My sleep cycle is a lot more balanced with remote work
ix. I have been able to find time to pick up a hobby
x. I spend more time with family and better connect with them
xi. My stress levels have reduced working remote
xii. I have been able to commit more time to self-care
xiii. It is easy to adjust to a purely online setting to work
xiv. My home environment allows me to focus on work

(Questions **xv.** to **xxii.** are answered with options: Yes, No)

xv. Zero travel time to my work appeals to me
xvi. I like being close to my loved ones all day
xvii. I prefer not having to do much physical work
xviii. I like not having to interact with my colleagues in person
xix. I dislike not being able to go outside and exercise
xx. I don't like not being able to travel
xxi. I dislike the effect that remote work has on my social life
xxii. Overall, I prefer: (Remote Work, Work in-person)

Compilation of References

Abdullah, A., & Chemmangat, K. (2020). A Computationally Efficient sEMG based Silent Speech Interface using Channel Reduction and Decision Tree based Classification. Elsevier.

Acharya, J., Chuadhary, A., Chhabria, A., & Jangale, S. (2021). Detecting Malware, Malicious URLs and Virus Using Machine Learning and Signature Matching. *2021 2nd International Conference for Emerging Technology (INCET)*, 1-5. 10.1109/INCET51464.2021.9456440

Acharya, U. R., Oh, S. L., Hagiwara, Y., Tan, J. H., Adeli, H., & Subha, D. P. (2018). Automated EEG-based screening of depression using deep convolutional neural network. *Computer Methods and Programs in Biomedicine*, *161*, 103–113. doi:10.1016/j.cmpb.2018.04.012 PMID:29852953

Affective Computing and Autism. (2006). *Massachusetts Institute of Technology*.

Agarwal, A., & Lavie, A. (2008). Meteor, m-bleu and m-ter: Evaluation metrics for highcorrelation with human rankings of machine translation output. In *Proceedings of the Third Workshop on Statistical Machine Translation*. Association for Computational Linguistics. 10.3115/1626394.1626406

Agrawal, U., Giripunje, S., & Bajaj, P. (2013). Emotion and gesture recognition with soft computing tool for drivers assistance system in human centered transportation. *Proceedings of the 2013 IEEE International Conference on Systems, Man, and Cybernetics (SMC'13)*, 4612-4616. 10.1109/SMC.2013.785

Alarcao, S. M., & Fonseca, M. J. (2019). Emotions Recognition Using EEG Signals: A Survey. *IEEE Transactions on Affective Computing*, *10*(3), 374–393. doi:10.1109/TAFFC.2017.2714671

Alhazzani, N. (2020). MOOC's impact on higher education. *Social Sciences & Humanities Open*, *2*(1), 100030. doi:10.1016/j.ssaho.2020.100030 PMID:34171022

Almeida, A., & Azkune, G. (2018, February). Predicting Human Behaviour with Recurrent Neural Networks. *Applied Sciences (Basel, Switzerland)*, *8*(2), 305. doi:10.3390/app8020305

Althuwaynee, O. F., Pradhan, B., Park, H. J., & Lee, J. H. (2014). A novel ensemble decision tree-based CHi-squared Automatic Interaction Detection (CHAID) and multivariate logistic regression models in landslide susceptibility mapping. *Landslides*, *11*(6), 1063–1078. doi:10.100710346-014-0466-0

Asthana, P., & Gupta, V. K. (2019). Role of Artificial Intelligence in Dealing with Emotional and Behavioural Disorders. *Journal*, *9*(1). http://www. ijmra. us

Ay, B., Yildirim, O., Talo, M., Baloglu, U. B., Aydin, G., Puthankattil, S. D., & Acharya, U. R. (2019). Automated Depression Detection Using Deep Representation and Sequence Learning with EEG Signals. *Journal of Medical Systems*, *43*(7), 205. doi:10.100710916-019-1345-y PMID:31139932

Baidya, S., & Levorato, M. (2018, September). Content-Aware Cognitive Interference Control for Urban IoT Systems. *IEEE Transactions on Cognitive Communications and Networking*, *4*(3), 500–512. doi:10.1109/TCCN.2018.2815604

Bai, W., Quan, C., & Luo, Z. (2018, February). Uncertainty Flow Facilitates Zero-Shot Multi-Label Learning in Affective Facial Analysis. *Applied Sciences (Basel, Switzerland)*, *8*(2), 300. doi:10.3390/app8020300

Bandura, A (1986), The Explanatory and Predictive Scope of Self-Efficacy Theory. *Journal of Social and Clinical Psychology, 4.*

Bandura, A. (1978). Reflections on self-efficacy. *Advances in Behaviour Research and Therapy, 1*(4), 237-269.

Baraka, K., Marta Couto Francisco, S. M., & Paiva, A. (2022). *Sequencing Matters*. Investigating Suitable Action Sequences in Robot-Assisted Autism Therapy Frontiers in Robotics and AI, doi:10.3389/frobt.2022.784249

Bazmara, M., Movahed, S. V., & Ramadhani, S. (2013). KNN Algorithm for Consulting Behavioral Disorders in Children. *Journal of Basic and Applied Scientific Research*, *3*, 12.

Benjamin, Heffetz, & Kimball. (2014). Beyond Happiness and Satisfaction: Toward Well-being Indces based on Stated Preference. *American Economic Review, 104*(9), 2698-2735.

Beverly, W. W. B. I. A., Dragon, T., & David, C. R. P. (2009). Affect-aware tutors: Recognising and responding to student affect. *International Journal of Learning Technology*, *4*(3/4), 129–164. doi:10.1504/IJLT.2009.028804

Bird, J. J., Ekart, A., & Faria, D. R. (2019). *Mental Emotional Sentiment Classification with an EEG-based Brain-machine Interface HANDLE Project (EU FP7) View project EMG-controlled 3D Printed Prosthetic Hand for Academia View project*. Academic Press.

Bird, J. J., Faria, D. R., Manso, L. J., Ekárt, A., & Buckingham, C. D. (2019). A deep evolutionary approach to bioinspired classifier optimisation for brain-machine interaction. *Complexity*, *2019*, 1–14. Advance online publication. doi:10.1155/2019/4316548

Boril, H., Kleinschmidt, T., Boyraz, P., & Hansen, J. H. L. (2010). Impact of cognitive load and frustration on drivers' speech. *The Journal of the Acoustical Society of America, 127.* Advance online publication. doi:10.1121/1.3385171

Bos, D. O. (2006). EEG-based emotion recognition - The Influence of Visual and Auditory Stimuli. *Capita Selecta*, 1–17.

Boser,, Guyon, & Vapnik. (1992). A training algorithm for optimal margin classifiers. *Proceedings of the fifth Annual Workshop on Computational Learning Theory*, 144–152.

Bosseler, A., & Massaro, D. W. (2004). Development and Evaluation of a Computer-Animated Tutor for Vocabulary and Language Learning in Children with Autism. *Journal of Autism and Developmental Disorders*, *33*(6), 653–672. doi:10.1023/B:JADD.0000006002.82367.4f PMID:14714934

Bunluechokchai, C., & Leeudomwong, T. (2010). Discrete Wavelet Transform -based Baseline Wandering Removal for High Resolution Electrocardiogram. *Int J Applied Biomed Eng, 3*.

Byeon, H. (2020). Development of a depression in Parkinson's disease prediction model using machine learning. *World Journal of Psychiatry*, *10*(10), 234–244. doi:10.5498/wjp.v10.i10.234 PMID:33134114

Chaidi & Drigas. (2020). Autism, Expression, and Understanding of Emotions: Literature Review Article. *International Journal of Online Engineering*, 125–147.

Chauhan, J., & Goel, A. (2017). An Overview of MOOC in India Education Management View project An Overview of MOOC in India. *International Journal of Computer Trends and Technology*, *49*(2), 111–120. Advance online publication. doi:10.14445/22312803/IJCTT-V49P117

Chebib, K. (2020). *IoT applications in the fight against COVID-19.* https://www.gsma.com/mobilefordevelopment/blog/iot-applications-in-the-fight-against-covid-19

Chen, Yang, Hao, Mao, & Kai. (2017). A 5G cognitive system for healthcare. *Big Data Cognit. Comput.*

Chen, M. Y. (2011). Predicting corporate financial distress based on integration of decision tree classification and logistic regression. *Expert Systems with Applications*, *38*(9), 11261–11272. doi:10.1016/j.eswa.2011.02.173

Chen, M., Hao, Y., Hu, L., Huang, K., & Lau, V. (2017, December). Green and mobility aware caching in 5G networks. *IEEE Transactions on Wireless Communications*, *6*(2), 8347–8836. doi:10.1109/TWC.2017.2760830

Chen, M., Herrera, F., & Hwang, A. K. (2018, April). Cognitive computing: Architecture, Technologies and Intelligent Applications. *IEEE Access: Practical Innovations, Open Solutions*, *6*, 19774–19783. doi:10.1109/ACCESS.2018.2791469

Chen, M., Miao, Y., Hao, Y., & Huang, K. (2017). *Narrow band Internet of Things* (Vol. 5). IEEE Access.

Chen, X., & Lawrence Zitnick, C. (2015). Mind's eye: A recurrent visual representation for image caption generation. *Proceedings of the IEEE Conference on Computer Vision and Pattern Recognition*, 2422–2431. 10.1109/CVPR.2015.7298856

Chen, Y., Argentinis, J. E., & Weber, G. (2016). *IBM Watson: How cognitive computing can be applied to big data challenges in life sciences research* (Vol. 38). Clin Therapeutics.

Cicchetti, D., & Sroufe, L. A. (1976). The relationship between affective and cognitive development in Down's syndrome infants. *Child Development*, *47*(4), 920–929. doi:10.2307/1128427 PMID:137105

Clayton, J., Wellington, S., Valentini-Botinhao, C., & Watts, O. (2019). *Decoding imagined, heard, and spoken speech: classification and regression of EEG using a 14-channel dry-contact mobile headset*. The University of Edinburgh.

Cohen, I. L., & Flory, M. J. (2019). Autism spectrum disorder decision tree subgroups predict adaptive behavior and autism severity trajectories in children with ASD. *Journal of Autism and Developmental Disorders*, *49*(4), 1423–1437. doi:10.100710803-018-3830-4 PMID:30511124

Cohn, J., & Kanade, T. (2007). Automated facial image analysis for measurement of emotion expression. In J. A. Coan & J. B. Allen (Eds.), The handbook of emotion elicitation and assessment (pp. 222–238). Oxford.

Comstock (1984). *Traps for the young*. Academic Press.

Conroy, D. A., Hadler, N. L., & Cho, E. (2021). The effects of COVID-19 stay-at-home order on sleep, health, and working patterns: A survey study of US health workers. *Journal of Clinical Sleep Medicine*, *17*(2), 185–191.

Crane, R. A., & Comley, S. (2021). Influence of social learning on the completion rate of massive online open courses. *Education and Information Technologies*, *26*(2), 2285–2293. doi:10.100710639-020-10362-6

Crews, S. D., Bender, H., Cook, C. R., Gresham, F. M., Kern, L., & Vanderwood, M. (2007). Risk and protective factors of emotional and/or behavioral disorders in children and adolescents: A mega-analytic synthesis. *Behavioral Disorders*, *32*(2), 64–77. doi:10.1177/019874290703200201

Cruz, A. C., & Rinaldi, A. (2017). Video summarization for expression analysis of motor vehicle operators. *Proceedings of the International Conference on Universal Access in Human-Computer Interaction*, 313-323. 10.1007/978-3-319-58706-6_25

D'Mello, S., & Graesser, A. (2011). The Half-Life of Cognitive-Affective States during Complex Learning. *Cognition and Emotion*, *25*(7), 1299–1308. doi:10.1080/02699931.2011.613668 PMID:21942577

Dalal, N., & Triggs, B. (2005). Histograms of oriented gradients for human detection. *IEEE Conference on Computer Vision and Pattern Recognition, 1*, 886–893. 10.1109/CVPR.2005.177

Danielson, M. L., Bitsko, R. H., Ghandour, R. M., Holbrook, J. R., Kogan, M. D., & Blumberg, S. J. (2018). Prevalence of parent-reported ADHD diagnosis and associated treatment among US children and adolescents, 2016. *Journal of Clinical Child and Adolescent Psychology*, *47*(2), 199–212. doi:10.1080/15374416.2017.1417860 PMID:29363986

Dash, D., Ferrari, P., & Wang, J. (2020). Decoding Imagined and Spoken Phrases from Non-invasive Neural (MEG) Signals. *Frontiers in Neuroscience*.

Davoli, L., Martalò, M., Cilfone, A., Belli, L., Ferrari, G., Presta, R., Montanari, R., Mengoni, M., Giraldi, L., Amparore, E. G., Botta, M., Drago, I., Carbonara, G., Castellano, A., & Plomp, J. (2020, December). On Driver Behavior Recognition for Increased Safety: A Roadmap. *Safety (Basel, Switzerland)*, *6*(4), 55. doi:10.3390afety6040055

De Marneffe, MacCartney, & Manning. (2006). Generating typed dependency parses from phrase structure parses. *Proceedings of LREC, 6*, 449–454.

De', R., Pandey, N., & Pal, A. (2020). Impact of digital surge during Covid-19 pandemic: A viewpoint on research and practice. *International Journal of Information Management*, *55*, 102171. doi:10.1016/j.ijinfomgt.2020.102171

Diehl, Schmitt, Villano, & Crowell. (2012). The Clinical Use of Robots for Individuals with Autism Spectrum Disorders. *Critical Review*. Advance online publication. doi:10.1016/j.rasd.2011.05.006

Ding, D., Gebel, K., Phongsavan, P., Bauman, A. E., & Merom, D. (2014, June). Driving: A road to unhealthy lifestyles and poor health outcomes. *PLoS One*, *9*(6), 1–5. doi:10.1371/journal.pone.0094602 PMID:24911017

Ding, H., Ghazilla, R. A. R., Singh, R. S. K., & Wei, L. (2022, February). Deep learning method for risk identification under multiple physiological signals and PAD model. *Microprocessors and Microsystems*, *88*, 104393. doi:10.1016/j.micpro.2021.104393

D'Mello & Graesser. (2012). AutoTutor and affective AutoTutor: Learning by talking with cognitively and emotionally intelligent computers that talk back. *The ACM Transactions on Interactive Intelligent Systems*.

Dodia, S., Reddy Edla, D., Bablani, A., Ramesh, D., & Kuppili, V. (2019). An Efficient EEG based Deceit Identification Test using Wavelet Packet Transform and Linear Discriminant Analysis. *Journal of Neuroscience Methods*, *314*, 1–33. doi:10.1016/j.jneumeth.2019.01.007 PMID:30660481

Dormann, C., & Zapf, D. (2004). Customer-Related Social Stressors and Burnout. *Journal of Occupational Health Psychology*, *9*(1), 61–82.

Dulhare & Rasool. (2019). IoT Evolution and Security Challenges in Cyber Space: IoT Security. In *Countering Cyber Attacks and Preserving the Integrity and Availability of Critical Systems*. IGI Global.

Dulhare, U. N., & Rasool, S. (2022). Smart Airport System to Counter COVID-19 and Future Sustainability. In C. Satyanarayana, X. Z. Gao, C. Y. Ting, & N. B. Muppalaneni (Eds.), *Machine Learning and Internet of Things for Societal Issues. Advanced Technologies and Societal Change*. Springer. doi:10.1007/978-981-16-5090-1_5

Ekman, P. (1999). *Basic emotions. In Handbook of cognition and emotion*. Wiley.

Eyben, F., Wöllmer, M., & Schuller, B. (2010). OpenSMILE: The Munich versatile and fast open-source audio feature extractor. *Proceedings of ACM Multimedia*, 1459-1462. DOI:10.1145/1873951.1874246

FitzGerald, N., Artzi, Y., & Zettlemoyer, L. (2013). Learning distributions over logical forms for referring expression generation. *Proceedings of the 2013 Conference on Empirical Methods in Natural Language Processing*, 1914–1925.

Forness, S. R., Freeman, S. F., Paparella, T., Kauffman, J. M., & Walker, H. M. (2012). Special education implications of point and cumulative prevalence for children with emotional or behavioral disorders. *Journal of Emotional and Behavioral Disorders*, *20*(1), 4–18. doi:10.1177/1063426611401624

Fossum Færevaag, E. (2021). *IoT's Evolution During COVID-19 and Into the 'New Normal'*. Channelfutures. https://www.channelfutures.com/iot/iots-evolution-during-covid-19-and-into-the-new-normal

Founoun & Hayar. (2018). Evaluation of the concept of the smart city through local regulation and the importance of local initiative. *2018 IEEE International Smart Cities Conference (ISC2)*, 1-6. 10.1109/ISC2.2018.8656933

Galanti, T., Guidetti, G., Mazzei, E., Zappalà, S., & Toscano, F. (2021). Work From Home During the COVID-19 Outbreak: The Impact on Employees' Remote Work Productivity, Engagement, and Stress. *Journal of Occupational and Environmental Medicine*, *63*(7), e426–e432. doi:10.1097/JOM.0000000000002236

Garber, M. C. (2017). Exercise as a Stress Coping Mechanism in a Pharmacy Student Population. *American Journal of Pharmaceutical Education*, *81*(3), 50.

García, E., Romero, C., Ventura, S., & De Castro, C. (2011). A collaborative educational association rule mining tool. *The Internet and Higher Education*, *14*(2), 77–88. doi:10.1016/j.iheduc.2010.07.006

Garcia-Garcia, Penichet, Lozano, & Fernando. (2021). *Using emotion recognition technologies to teach children with autism spectrum disorder how to identify and express emotions*. Academic Press.

Ghazanfar, M., & Prugel-Bennett, A. (2010). *An improved switching hybrid recommender system using Naive Bayes classifier and collaborative filtering*. Academic Press.

Gibbs, M., Mengel, F., & Siemroth, C. (2021). *Work from Home & Productivity: Evidence from Personnel & Analytics Data on IT Professionals*. University of Chicago, Becker Friedman Institute for Economics Working Work No. 2021-56. Available at SSRN: https://ssrn.com/abstract=3843197

Girshick, R. (2015). Fast r-cnn. *Proceedings of the IEEE International Conference on Computer Vision*, 1440–1448.

Golan, O., Ashwin, E., Granader, Y., McClintock, S., Day, K., Leggett, V., & Baron-Cohen, S. (2010). Enhancing emotion recognition in children with autism spectrum conditions: An intervention using animated vehicles with real emotional faces. *Journal of Autism and Developmental Disorders*, *40*(3), 269–279. doi:10.100710803-009-0862-9 PMID:19763807

Goldmen, E. (2012). *Living my life* (Vol. 1). Dover Publications, Inc.

Golland, D., Liang, P., & Klein, D. (2010). A game-theoretic approach to generating spatial descriptions. In *Proceedings of the 2010 Conference on Empirical Methods in Natural Language Processing*. Association for Computational Linguistics.

Gould, E., & Kassa, M. (2020). *Young workers hit hard by the COVID-19 economy*. Economic Policy Institute. https://www.epi.org/publication/young-workers-covid-recession/

Gouribhatla, R., & Pulugurtha, S. S. (2022, March). Drivers' behavior when driving vehicles with or without advanced driver assistance systems: A driver simulator-based study. *Transportation Research Interdisciplinary Perspectives, 13*, 100545. doi:10.1016/j.trip.2022.100545

Graesser, A., & D'Mello, S. K. (2011). Theoretical perspectives on affect and deep learning. In R. A. Calvo & S. K. D'Mello (Eds.), *New perspectives on affect and learning technologies* (Vol. 3, pp. 11–21). Springer. doi:10.1007/978-1-4419-9625-1_2

Grimm, M., Kroschel, K., Harris, H., Nass, C., Schuller, B. B., Rigoll, G., & Moosmayr, T. (2007). On the necessity and feasibility of detecting a driver's emotional state while driving. *Affective Computing and Intelligent Interaction, 4738*, 126–138. doi:10.1007/978-3-540-74889-2_12

Gu, X., Cao, Z., Jolfaei, A., Xu, P., Wu, D., Jung, T. P., & Lin, C. T. (2020). EEG-based Brain-Computer Interfaces (BCIs): A Survey of Recent Studies on Signal Sensing Technologies and Computational Intelligence Approaches and their Applications. arXiv preprint arXiv:2001.11337

Guo, Yüce, & Thiran. (2014). Detecting emotional stress from facial expressions for driving safety. *Proceedings of the IEEE International Conference on Image Processing (ICIP'14), 1*, 5961-5965. 10.1109/ICIP.2014.7026203

Haag, A., Goronzy, S., Schaich, P., & Williams, J. (2004). Emotion Recognition Using Bio-sensors: First Steps towards an Automatic System. *Affective Dialogue Systems, i*, 36–48. doi:10.1007/978-3-540-24842-2_4

Hamann, D. L., & Gordon, D. G. (2000). Burnout An Occupational Hazard. *Music Educators Journal, 87*(3).

Hamann, K., Glazier, R. A., Wilson, B. M., & Pollock, P. H. (2021). Online teaching, student success, and retention in political science courses. *European Political Science, 20*(3), 427–439. doi:10.105741304-020-00282-x

Harry, M. (2021). *Americal on Films, Representing Race, Class, Gender, and Sexuality at the Movies* (3rd ed.). Blackwell Publishing Ltd., John Wiley & Sons.

Hasan, Kaur, Hasan, & Feng. (2019). Cognitive Internet of Vehicles: Motivation, Layered Architecture and Security Issues. *International Conference on Sustainable Technologies for Industry 4.0*.

Hemanth, D. J. (2020). EEG signal based Modified Kohonen Neural Networks for Classification of Human Mental Emotions. *Journal of Artificial Intelligence and Systems, 2*(1), 1–13. doi:10.33969/AIS.2020.21001

Hiraki, H., & Rekimoto, J. (2021). SilentMask: Mask-type Silent Speech Interface with Measurement of Mouth Movement. Association for Computing Machinery.

Hochreiter, S., & Schmidhuber, J. (1997). Long short-term memory. *Neural Computation, 9*(8), 1735–1780. doi:10.1162/neco.1997.9.8.1735 PMID:9377276

Hodosh, M., Young, P., & Hockenmaier, J. (2013). Framing image description as a ranking task: Data, models and evaluation metrics. *Journal of Artificial Intelligence Research, 47*, 853–899. doi:10.1613/jair.3994

Hökkä, P., Vähäsantanen, K., & Paloniemi, S. (2020). Emotions in Learning at Work: A Literature Review. *Vocations and Learning, 13*(1), 1–25. doi:10.100712186-019-09226-z

Holdnack & Saklofske. (2019). WISC-V and the Personalized Assessment Approach. WISC-V.

Hurford, R., Martin, A., & Larsen, P. (2006). Designing Wearables. *2006 10th IEEE International Symposium on Wearable Computers,* 133-134. 10.1109/ISWC.2006.286362

Imani, M., & Montazer, G. A. (2019). A survey of emotion recognition methods with emphasis on E-Learning environments. *Journal of Network and Computer Applications, 147*(April), 102423. doi:10.1016/j.jnca.2019.102423

Jeon, M. (2016). Don't cry while you're driving: Sad driving is as bad as angry driving. *International Journal of Human-Computer Interaction, 32*(10), 777–790. doi:10.1080/10447318.2016.1198524

Jeon, M., Roberts, J., Raman, P., Yim, J.-B., & Walker, B. N. (2011). Participatory design process for an in-vehicle affect detection and regulation system for various drivers. *Proceedings of the 13th International ACM SIGACCESS Conference on Computers and Accessibility,* 271-272. 10.1145/2049536.2049602

Jerritta, S., Murugappan, M., Nagarajan, R., & Wan, K. (2011). Physiological signals based human emotion Recognition: A review. *Signal Processing and Its Applications (CSPA), 2011 IEEE 7th International Colloquium On,* 410–415. 10.1109/CSPA.2011.5759912

Johnson, J., Karpathy, A., & Li, F.-F. (2016). Densecap: Fully convolutional localization networks for dense captioning. *Proceedings of the IEEE Conference on Computer Vision and Pattern Recognition,* 4565–4574. 10.1109/CVPR.2016.494

Jones, & Jonsson. (2008). Using paralinguistic cues in speech to recognise emotions in older car drivers. In Lecture Notes in Computer Science: Vol. 4868. *Affect and Emotion in Human-Computer Interaction* (pp. 229–240). Springer. doi:10.1007/978-3-540-85099-1_20

Jonsson, U., Choque Olsson, N., & Bölte, S. (2016). Can findings from randomized controlled trials of social skills training in autism spectrum disorder be generalized? The neglected dimension of external validity. *Autism, 20*(3), 295–305. doi:10.1177/1362361315583817 PMID:25964654

Karen, R. (1992, February). Sharme. *Atlantic Monthly,* 40–70.

Karpathy, A., Joulin, A., & Fei F Li, F. (2014). *Deep fragment embeddings for bidirectional image sentence mapping*. Advances in Neural Information Processing systems.

Kazemzadeh, Ordonez, Matten, & Berg. (2014). ReferItGame: Referring to Objects in Photographs of Natural Scenes. *Proceedings of the Conference on Empirical Methods in Natural Language Processing (EMNLP)*, 787–798.

Khateeb, M., Anwar, S. M., & Alnowami, M. (2021). Multi-Domain Feature Fusion for Emotion Classification Using DEAP Dataset. *IEEE Access: Practical Innovations, Open Solutions*, 9, 12134–12142. doi:10.1109/ACCESS.2021.3051281

Kim, J. Y., Lee, H., Son, J., & Park, J. (2015). Smart home web of objects-based IoT management model and methods for home data mining. *2015 17th Asia-Pacific Network Operations and Management Symposium (APNOMS)*, 327-331. 10.1109/APNOMS.2015.7275448

Kim, M., Hong, J., & Man Ro, Y. (2021). Lip to Speech Synthesis with Visual Context Attentional GAN. *35th Conference on Neural Information Processing Systems (NeurIPS 2021)*, 1-13.

Kim, Y., & Choi, A. (2020). EEG-Based Emotion Classification Using Long Short-Term Memory Network with Attention Mechanism. *Sensors (Basel)*, *20*(23), 6727. doi:10.339020236727 PMID:33255539

King, G., & Zeng, L. (2001). Logistic regression in rare events data. *Political Analysis*, *9*(2), 137–163.

Kiros, R., Salakhutdinov, R., & Zemel, R. (2014). Multimodal neural language models. *Proceedings of the 31st International Conference on Machine Learning (ICML-14)*, 595–603.

Koctúrová, M., & Juhár, J. (2018, August). An overview of BCI-based speech recognition methods. In *2018 World Symposium on Digital Intelligence for Systems and Machines (DISA)* (pp. 327-330). IEEE. 10.1109/DISA.2018.8490536

Kolli, A., Fasih, A., Al Machot, F., & Kyamakya, K. (2011). Non-intrusive car driver's emotion recognition using thermal camera. In *Proceedings of the 3rd International Workshop on Nonlinear Dynamics and Synchronization (INDS'11) and the 16th International Symposium on Theoretical Electrical Engineering (ISTET'11)*. IEEE. 10.1109/INDS.2011.6024802

Kort, B., Reilly, R., & Picard, R. W. (2001). *An affective model of interplay between emotions and learning: Reengineering educational pedagogy—Building a learning companion*. Paper presented at the International Conference on Advanced Learning Technologies, Madison, WI. 10.1109/ICALT.2001.943850

Krishna, G., Han, Y., Tran, C., Carnahan, M., & Tewfik, A. (2020). *State-of-the-art Speech Recognition using EEG and Towards Decoding of Speech Spectrum from EEG*. arxiv: 1908.05743v5.

Kumar, S., Shanker, R., & Verma, S. (2018). Context Aware Dynamic Permission Model: A Retrospect of Privacy and Security in Android System. *2018 International Conference on Intelligent Circuits and Systems (ICICS)*, 324-329. 10.1109/ICICS.2018.00073

Lange, K. W., Reichl, S., Lange, K. M., Tucha, L., & Tucha, O. (2010). The history of attention deficit hyperactivity disorder. *Attention Deficit and Hyperactivity Disorders*, 2(4), 241–255. doi:10.100712402-010-0045-8 PMID:21258430

LeCun, Y., Bottou, L., Bengio, Y., & Haffner, P. (1998). Gradient-based learning applied to document recognition". *Proceedings of the IEEE*, 86(11), 2278–2324. doi:10.1109/5.726791

Liaw, A., & Wiener, M. (2002, December). Classification and regression by randomForest. *R News*, 2, 18–22.

Li, H., Han, J., Li, S., Wang, H., Xiang, H., & Wang, X. (2022, January). Abnormal Driving Behavior Recognition Method Based on Smart Phone Sensor and CNN-LSTM. *International Journal of Science and Engineering Applications*, 11(1), 1–8. doi:10.7753/IJSEA1101.1001

Lindhiem, O., Bennett, C. B., Hipwell, A. E., & Pardini, D. A. (2015). Beyond symptom counts for diagnosing oppositional defiant disorder and conduct disorder? *Journal of Abnormal Child Psychology*, 43(7), 1379–1387. doi:10.100710802-015-0007-x PMID:25788042

Liu, G. D., Li, Y. C., Zhang, W., & Zhang, L. (2020). A brief review of artificial intelligence applications and algorithms for psychiatric disorders. *Engineering*, 6(4), 462–467. doi:10.1016/j.eng.2019.06.008

Liu, Y., Sourina, O., & Nguyen, M. K. (2010). Real-Time EEG-Based Human Emotion Recognition and Visualization. *2010 International Conference on Cyberworlds*, 262–269. 10.1109/CW.2010.37

Li, Y., Zheng, W., Wang, L., Zong, Y., & Cui, Z. (2019). From Regional to Global Brain: A Novel Hierarchical Spatial-Temporal Neural Network Model for EEG Emotion Recognition. *IEEE Transactions on Affective Computing*, 1–1. doi:10.1109/TAFFC.2019.2922912

Lowe, D. (2004). Distinctive image features from scale-invariant keypoints. *International Journal of Computer Vision*, 60(2), 91–110. doi:10.1023/B:VISI.0000029664.99615.94

Luo, J., Wang, J., Cheng, N., Jiang, G., & Xiao, J. (2020). *End–to–End Silent Speech Recognition with Acoustic Sensing*. arxiv: 2011.11315v1.

Ma, P., Martinez, B., Petridis, S., & Pantic, M. (2021). Towards Practical Lipreading with Distilled and Efficient Models. doi:10.1109/ICASSP39728.2021.9415063

Ma, S., Jin, D., Zhang, M., Zhang, B., Wang, Y., Li, G., & Yang, M. (2019). Silent Speech Recognition Based on Surface Electromyography. IEEE.

Mansoor, A., Usman, M., Jamil, N., & Naeem, A. (2020). Deep Learning Algorithm for Brain-Computer Interface. *Scientific Programming*, 2020, 1–12. doi:10.1155/2020/5762149

Mao, J., Huang, J., Toshev, A., & Camburu, O. (2016). Generation and comprehension of unambiguous object descriptions. *Proceedings of the IEEE Conference on Computer Vision and Pattern Recognition*, 11–20.

Mao, J., Xu, W., Yang, Y., Wang, J., Huang, Z., & Yuille, A. (2015). Deep captioning with multimodal recurrent neural networks (m-rnn). *International Conference on Learning Representations (ICLR)*.

Mao, X., & Li, Z. (2010). Agent based affective tutoring systems: A pilot study. *Computers & Education, 55*(1), 202–208. doi:10.1016/j.compedu.2010.01.005

Matthews, K. (2018). *5 Challenges Facing Health Care IoT in 2019*. Iotforall. https://www.iotforall.com/5-challenges-facing-iot-healthcare-2019

Ma, Z., Mahmoud, M., Robinson, P., Dias, E., & Skrypchuk, L. (2017). Automatic detection of a driver's complex mental states. *Proceedings of the International Conference on Computational Science and Its Applications*, 678-691. 10.1007/978-3-319-62398-6_48

Mazumder, D. (2021). A novel approach to IoT based health status monitoring of COVID-19 patient. *2021 International Conference on Science & Contemporary Technologies (ICSCT)*, 1-4. 10.1109/ICSCT53883.2021.9642608

Michelini, G., Kitsune, G. L., Cheung, C. H., Brandeis, D., Banaschewski, T., Asherson, P., & Kuntsi, J. (2016). Attention-deficit/hyperactivity disorder remission is linked to better neurophysiological error detection and attention-vigilance processes. *Biological Psychiatry, 80*(12), 923–932. doi:10.1016/j.biopsych.2016.06.021 PMID:27591125

Min, H., & Nasir, M. K. M. (2020). Self-Regulated Learning In A Massive Open Online Course: A Review of Literature. *European Journal of Interactive Multimedia and Education, 1*(2), e02007. doi:10.30935/ejimed/8403

Mitchell, M., van Deemter, K., & Reiter, E. (2010). Natural reference to objects in a visual domain. In *Proceedings of the 6th International Natural Language Generation Conference*. Association for Computational Linguistics.

Mitchell, M., Van Deemter, K., & Reiter, E. (2013). *Generating Expressions that Refer to Visible Objects*. HLT-NAACL.

Molina, G. G., Tsoneva, T., & Nijholt, A. (2009). Emotional brain-computer interfaces. *2009 3rd International Conference on Affective Computing and Intelligent Interaction and Workshops*, 1–9. 10.1109/ACII.2009.5349478

Morelli, M., Cattelino, E., Baiocco, R., Trumello, C., Babore, A., Candelori, C., & Chirumbolo, A. (2020). Parents and Children During the COVID-19 Lockdown: The Influence of Parenting Distress and Parenting Self-Efficacy on Children's Emotional Well-Being. *Frontiers in Psychology, 11*, 584645. doi:10.3389/fpsyg.2020.584645

Moriyama, T. (2012). Face analysis of aggressive moods in automobile driving using mutual subspace method. *Proceedings of the 21st International Conference on Pattern Recognition*.

Morley, D., & Parker, C. (2013). *Understanding Computers: Today and Tomorrow*. Cengage Learning.

Morse, K. F., Fine, P. A., & Friedlander, K. J. (2021). Creativity and Leisure During COVID-19: Examining the Relationship Between Leisure Activities, Motivations, and Psychological Well-Being. *Frontiers in Psychology*, *12*, 609967. doi:10.3389/fpsyg.2021.609967

Mumtaz, W., & Qayyum, A. (2019). A deep learning framework for automatic diagnosis of unipolar depression. *International Journal of Medical Informatics*, *132*, 103983. doi:10.1016/j.ijmedinf.2019.103983 PMID:31586827

Muñoz, S., Sánchez, E., & Iglesias, C. A. (2020). An emotion-aware learning analytics system based on semantic task automation. *Electronics (Switzerland)*, *9*(8), 1–24. doi:10.3390/electronics9081194

N. K., & J. R. (2019, March). A Robust User Sentiment Biterm Topic Mixture Model Based on User Aggregation Strategy to Avoid Data Sparsity for Short Text. *Journal of Medical Systems*, *43*(4). PMID:30834466

Nelson, J. R., Stage, S., Duppong-Hurley, K., Synhorst, L., & Epstein, M. H. (2007). Risk factors predictive of the problem behavior of children at risk for emotional and behavioral disorders. *Exceptional Children*, *73*(3), 367–379. doi:10.1177/001440290707300306

Nimala, K., Jebakumar, R., & Saravanan, M. (2020, July). Sentiment topic sarcasm mixture model to distinguish sarcasm prevalent topics based on the sentiment bearing words in the tweets. *Journal of Ambient Intelligence and Humanized Computing*, *12*(6), 6801–6810. doi:10.100712652-020-02315-1

Nimala, K., Magesh, S., & Arasan, R. T. (2018). Hash tag based topic modelling techniques for twitter by tweet aggregation strategy. *Journal of Advanced Research in Dynamical and Control Systems*, *3*, 571–578.

Nisiforou, E. A., & Zaphiris, P. (2020). *Let me play: Unfolding the research landscape on ICT as a play-based tool for children with disabilities. Universal Access in the Information Society, 19(1).*

Noori, R., Karbassi, A. R., Moghaddamnia, A., Han, D., Zokaei-Ashtiani, M. H., Farokhnia, A., & Gousheh, M. G. (2011). Assessment of input variables determination on the SVM model performance using PCA, Gamma test, and forward selection techniques for monthly stream flow prediction. *Journal of Hydrology (Amsterdam)*, *401*(3-4), 177–189. doi:10.1016/j.jhydrol.2011.02.021

Oehl, M., Siebert, F. W., Tews, T.-K., Höger, R., & Pfister, H.-R. (2011). Improving human-machine interaction: A non invasive approach to detect emotions in car drivers. In. Lecture Notes in Computer Science: Vol. 6763. *Human-Computer Interaction: Towards Mobile and Intelligent Interaction Environments* (pp. 577–585). Springer. doi:10.1007/978-3-642-21616-9_65

Ojala, T., Pietikäinen, M., & Mäenpää, T. (2000). Gray scale and rotation invariant texture classification with local binary patterns. In *European Conference on Computer Vision*. Springer. 10.1007/3-540-45054-8_27

Papineni, K., Roukos, S., Ward, T., & Zhu, W.-J. (2002). BLEU: A method for automatic evaluation of machine translation. In *Proceedings of the 40th Annual Meeting on Association for Computational Linguistics*. Association for Computational Linguistics.

Paschero, M., Del Vescovo, G., Benucci, L., Rizzi, A., Santello, M., Fabbri, G., & Frattale Mascioli, F. M. (2012). A real time classifier for emotion and stress recognition in a vehicle driver. *Proceedings of the IEEE International Symposium on Industrial Electronics*, 1690-1695. 10.1109/ISIE.2012.6237345

Peters, P., Tijdens, K. G., & Wetzels, C. (2004). 'Employees' opportunities, preferences, and practices in telecommuting adoption'. *Information & Management*, *41*(4), 469–482.

Phan, T.-D.-T., Kim, S.-H., Yang, H.-J., & Lee, G.-S. (2021). EEG-Based Emotion Recognition by Convolutional Neural Network with Multi-Scale Kernels. *Sensors (Basel)*, *21*(15), 5092. doi:10.339021155092 PMID:34372327

Philip, R. C. M., Whalley, H. C., Stanfield, A. C., Sprengelmeyer, R., Santos, I. M., Young, A. W., Atkinson, A. P., Calder, A. J., Johnstone, E. C., Lawrie, S. M., & Hall, J. (2010). Deficits in facial, body movement and vocal emotional processing in autism spectrum disorders. *Psychological Medicine*, *40*(11), 1919–1929. doi:10.1017/S0033291709992364 PMID:20102666

Picard, R. W., Vyzas, E., & Healey, J. (2001). Toward machine emotional intelligence: Analysis of affective\nphysiological state. *IEEE Transactions on Pattern Analysis and Machine Intelligence*, *23*(10), 1175–1191. doi:10.1109/34.954607

Pichard, R. W. (2009). Future affective technology for autism and emotion communication. *USA Phil. Trans. R. Soc. B*, *364*, 3575–3584. doi:10.1098/rstb.2009.0143

Pinkham & Badcock. (2020). Assessing Cognition and Social Cognition in Schizophrenia & Related Disorders. A Clinical Introduction to Psychosis, 201-225.

Pisano, S., Muratori, P., Gorga, C., Levantini, V., Iuliano, R., Catone, G., & Masi, G. (2017). Conduct disorders and psychopathy in children and adolescents: Aetiology, clinical presentation and treatment strategies of callous-unemotional traits. *Italian Journal of Pediatrics*, *43*(1), 1–11. doi:10.118613052-017-0404-6 PMID:28931400

Provost, E. M., Black, M., & Flôres, E. L. (2011). Rachel: Design of an emotionally targeted interactive agent for children with autism. *IEEE International Conference on Multimedia and Expo (ICME)*.

Rabbani, Q., Milsap, G., & Crone, N. E. (2019). The potential for a speech brain–computer interface using chronic electrocorticography. *Neurotherapeutics; the Journal of the American Society for Experimental NeuroTherapeutics*, *16*(1), 144–165. doi:10.100713311-018-00692-2 PMID:30617653

Radhamani, E., & Krishnaveni, K. (2016). Diagnosis and Evaluation of ADHD using MLP and SVM Classifiers. *Indian Journal of Science and Technology*, *9*(19), 93853. doi:10.17485/ijst/2016/v9i19/93853

Räikkönen, K., Gissler, M., & Kajantie, E. (2020). Associations between maternal antenatal corticosteroid treatment and mental and behavioral disorders in children. *Journal of the American Medical Association, 323*(19), 1924–1933. doi:10.1001/jama.2020.3937 PMID:32427304

Ramos-Aguiar, L. R., & Álvarez-Rodríguez, F. J. (2021). Teaching Emotions in Children With Autism Spectrum Disorder Through a Computer Program With Tangible Interfaces. IEEE Revista Iberoamericana, 16(4), 365 – 371.

Rangasamy, S., D'Mello, S. R., & Narayanan, V. (2013). *Epigenetics, Autism Spectrum, and Neurodevelopmental Disorders. Journal of the American Society for Experimental NeuroTherapeutics, 10*(4). Advance online publication. doi:10.100713311-013-0227-0 PMID:24104594

Rasool & Dulhare. (2022). Data Center Security. In *Green Computing in Network Security: Energy Efficient Solutions for Business and Home*. CRC Press. doi:10.1201/9781003097198

Rauch, S. (2021). *Using IoT Applications to Manage the Spread of Covid-19*. Simplilearn. https://www.simplilearn.com/using-iot-applications-to-manage-the-spread-of-covid-19-article

Rehman, U., Shahnawaz, M. G., & Khan, N. H. (2021). Depression, Anxiety and Stress Among Indians in Times of Covid-19 Lockdown. *Community Mental Health Journal, 57*(1), 42–48. doi:10.100710597-020-00664-x

Rennie, Marcheret, Mroueh, Ross, & Goel. (2017). Self-critical sequence training for image captioning. *Proceedings of the IEEE Conference on Computer Vision and Pattern Recognition (CVPR)*, 1179–1195.

Ren, S., He, K., Girshick, R., & Sun, J. (2015). *Faster R-CNN: Towards real-time object detection with region proposal networks*. Advances in Neural Information Processing Systems.

Ren, Z., Wang, X., Zhang, N., Lv, X., & Li, L.-J. (2017). Deep Reinforcement Learningbased Image Captioning with Embedding Reward. *Proceedings of the IEEE Conference on Computer Vision and Pattern Recognition (CVPR)*, 1151–1159.

Russell, D., & Kane, J. (1977, November). Sulfide Stress Cracking of High-Strength Steels in Laboratory and Oilfield Environments. *Journal of Petroleum Technology, 29*(11), 1483–1488.

Saraswat, P., Garg, K., Tripathi, R., & Agarwal, A. (2019). Encryption Algorithm Based on Neural Network. *2019 4th International Conference on Internet of Things: Smart Innovation and Usages (IoT-SIU)*, 1-5. 10.1109/IoT-SIU.2019.8777637

Sarkar, A., Banerjee, A., Singh, P. K., & Sarkar, R. (2021, September). 3D Human Action Recognition: Through the eyes of researchers. *Expert Systems with Applications, 193*.

Schuller, B., Lang, M., & Rigoll, G. (2006). Recognition of spontaneous emotions by speech within automotive environment. *Tagungsband Fortschritte der Akustik (DAGA'06)*, 57-58. http://www.mmk.ei.tum.de/publ/pdf/06/06sch5.pdf

Selvaraj, J., Murugappan, M., Wan, K., & Yaacob, S. (2013). Classification of emotional states from electrocardiogram signals: A non-linear approach based on Hurst. *Biomedical Engineering Online*, *12*(1), 44. doi:10.1186/1475-925X-12-44 PMID:23680041

Shah, D. (2020). *By The Numbers: MOOCs in 2020*. The Report by Class Central. https://www.classcentral.com/report/mooc-stats-2020/

Sharma, R., Pachori, R. B., & Sircar, P. (2020). Automated emotion recognition based on higher order statistics and deep learning algorithm. *Biomedical Signal Processing and Control*, *58*, 101867. doi:10.1016/j.bspc.2020.101867

Sharon, R., & Murthy, H. (2020). Correlation based Multi-phasal models for improved imagined speech EEG Recognition. doi:10.21437/SMM.2020-5

Shen, L., Wang, M., & Shen, R. (2009). Affective e-Learning: Using "emotional" data to improve learning in pervasive learning environment related work and the pervasive e-learning platform. *Journal of Educational Technology & Society*, *12*, 176–189.

Shu, L., Xie, J., Yang, M., Li, Z., Li, Z., Liao, D., Xu, X., & Yang, X. (2018). A Review of Emotion Recognition Using Physiological Signals. *Sensors (Basel)*, *18*(7), 2074. doi:10.339018072074 PMID:29958457

Siebert, F. W., Oehl, M., & Pfister, H.-R. (2010). The measurement of grip-strength in automobiles: A new approach to detect driver's emotions. In W. Karwowski & G. Salvendy (Eds.), *Advances in Human Factors, Ergonomics, and Safety in Manufacturing and Service Industry* (pp. 775–782). CRC Press. doi:10.1201/EBK1439834992-82

Simon. (2021). *The Internet of things (IoT). The Internet of things (IoT) describes....* https://simonbrard017.medium.com/the-internet-of-things-iot-describes-the-network-of-physical-objects-so-known-as-things-cb8c9c994603

Singh, A., Halgamuge, M. N., & Lakshmiganthan, R. (2017). *Impact of different data types on classifier performance of random forest, naive bayes, and k-nearest neighbors algorithms*. Academic Press.

Singhal, A., & Saxena, R. P. (2012). Software models for Smart Grid. *2012 First International Workshop on Software Engineering Challenges for the Smart Grid (SE-SmartGrids)*, 42-45. 10.1109/SE4SG.2012.6225717

Smola, A. J., & Schölkopf, B. (2004). A tutorial on support vector regression. *Statistics and Computing*, *14*, 199–222. https://doi.org/10.1023/B:STCO.0000035301.49549.88

Socher, R., Karpathy, A., & Quoc, V. (2014). Grounded compositional semantics for finding and describing images with sentences. *Transactions of the Association for Computational Linguistics*, *2*, 207–218. doi:10.1162/tacl_a_00177

Solomon, R. C. (2008). True to our Feelings: What our emotions are really telling us. Oxford University Press.

Song, T., Zheng, W., Lu, C., Zong, Y., Zhang, X., & Cui, Z. (2019). MPED: A Multi-Modal Physiological Emotion Database for Discrete Emotion Recognition. *IEEE Access: Practical Innovations, Open Solutions*, 7, 12177–12191. doi:10.1109/ACCESS.2019.2891579

Srividya, M., Mohanavalli, S., & Bhalaji, N. (2018). Behavioral modeling for mental health using machine learning algorithms. *Journal of Medical Systems*, 42(5), 1–12. doi:10.100710916-018-0934-5 PMID:29610979

Staples, D. S., Hulland, J. S., & Higgins, C. A. (1999). A Self-Efficacy Theory Explanation for the Management of Remote Workers in Virtual Organizations. *Organization Science*, 10(6), 758–776. https://dx.doi.org/10.1287/orsc.10.6.758

Stephan, M., Markus, S., & Gläser-Zikuda, M. (2019). Students' achievement emotions and online learning in teacher education. *Frontiers in Education*, 4(October), 1–12. doi:10.3389/feduc.2019.00109

Stress. (2018). In *Merriam-Webster.com*. https://www. merriam-webster.com

Sujatha, R., & Nimala, K. (2022). Text-based Conversation Analysis Techniques on Social Media using Statistical methods. *2022 International Conference on Advances in Computing, Communication and Applied Informatics (ACCAI)*. 10.1109/ACCAI53970.2022.9752562

Szentiványi, D., & Balázs, J. (2018). Quality of life in children and adolescents with symptoms or diagnosis of conduct disorder or oppositional defiant disorder. *Mental Health & Prevention*, 10, 1–8. doi:10.1016/j.mhp.2018.02.001

Tang, H., Xu, Y., Lin, A., Heidari, A. A., Wang, M., Chen, H., & Li, C. (2020). Predicting green consumption behaviors of students using efficient firefly grey wolf-assisted K-nearest neighbor classifiers. *IEEE Access: Practical Innovations, Open Solutions*, 8, 35546–35562. doi:10.1109/ACCESS.2020.2973763

Tavris, C. (1989). Anger: The misunderstood emotion. Simon & Schuster Inc.

Tawari, A., & Trivedi, M. M. (2010). Speech emotion analysis: Exploring the role of context. *IEEE Transactions on Multimedia*, 12(6), 502–509. doi:10.1109/TMM.2010.2058095

Tyng, C. M., Amin, H. U., Saad, M. N. M., & Malik, A. S. (2017). The influences of emotion on learning and memory. *Frontiers in Psychology*, 8(Aug), 1454. Advance online publication. doi:10.3389/fpsyg.2017.01454 PMID:28883804

Underwood, G., Chapman, P., Wright, S., & Crundall, D. (1999). Anger while driving. *Transportation Research Part F: Traffic Psychology and Behaviour*, 2(1), 55–68. doi:10.1016/S1369-8478(99)00006-6

van Deemter, K., van der Sluis, I., & Gatt, A. (2006). Building a semantically transparent corpus for the generation of referring expressions. In *Proceedings of the Fourth International Natural Language Generation Conference*. Association for Computational Linguistics. 10.3115/1706269.1706296

Viethen, J., & Dale, R. (2008). The use of spatial relations in referring expression generation. In *Proceedings of the Fifth International Natural Language Generation Conference.* Association for Computational Linguistics. 10.3115/1708322.1708334

Vinyals, O., Toshev, A., Bengio, S., & Erhan, D. (2015). Show and tell: A neural image caption generator. *Proceedings of the IEEE Conference on Computer Vision and Pattern Recognition,* 3156–3164. 10.1109/CVPR.2015.7298935

Vishwakarma, S. K., Upadhyaya, P., Kumari, B., & Mishra, A. K. (2019). Smart Energy Efficient Home Automation System Using IoT. *2019 4th International Conference on Internet of Things: Smart Innovation and Usages (IoT-SIU),* 1-4. 10.1109/IoT-SIU.2019.8777607

Vismara, L. A., & Rogers, S. J. (2010). Behavioral treatments in autism spectrum disorder: What do we know? *Annual Review of Clinical Psychology, 6*(1), 447–468. doi:10.1146/annurev.clinpsy.121208.131151 PMID:20192785

Wadman, R., Durkin, K., & Conti-Ramsden, G. (2011, June). Social Stress in Young People with specific language impairment. *Journal of Adolescence, 23*(3), 421–431.

Wang, M., Song, L., Yang, X., & Luo, C. (2016). A parallel-fusion RNN-LSTM architecture for image caption generation. In *2016 IEEE International Conference on Image Processing (ICIP).* IEEE. 10.1109/ICIP.2016.7533201

Warnecke, E., Quinn, S., Ogden, K., Towle, N., & Nelson, M. R. (2011). A randomised controlled trial of the effects of mindfulness practice on medical student stress levels. *Medical Education, 45,* 381–388.

Webb, G. I. (2011). Naïve Bayes. In C. Sammut & G. I. Webb (Eds.), Encyclopedia of Machine Learning. Springer. https://doi.org/10.1007/978-0-387-30164-8_576.

Wilford, E. (2022). *10 IoT technology trends to watch in 2022.* Iot-Analytics. https://iot-analytics.com/iot-technology-trends

WMA Declaration of Helsinki. (n.d.). https://www.wma.net/policies-post/wma-declaration-of-helsinki-ethical-principles-for-medical-research-involving-human-subjects/

Wood, R., & Bandura, A. (1989). Social Cognitive Theory of Organizational Management. *The Academy of Management Review, 14*(3), 361–384. www.jstor.org/stable/258173

Wood, W., & Eagly, A. H. (2002). A cross-cultural analysis of the behavior of women and men: Implications for the origins of sex differences. *Psychological Bulletin, 128*(5), 699–727.

Working from home during the COVID-19 lockdown: Changing preferences and the future of work. (n.d.). https://www.birmingham.ac.uk/Documents/college-social-sciences/business/research/wirc/epp-working-from-home-COVID-19-lockdown.pdf

World Health Organisation. (2020, March 11). *Coronavirus disease 2019 (COVID-19) Situation Report–51.* https://www.who.int/docs/default-source/coronaviruse/situation-reports/20200311-sitrep-51-covid-19.pdf?sfvrsn=1ba62e57_10

World Health Organization. (2020a). *Mental Health and Psychosocial Considerations During the COVID-19 Outbreak.* Available at: https://www.who.int/docs/default-source/coronaviruse/mental-health-considerations.pdf?sfvrsn=6d3578af_2

Wrobel, M. (2018, February). Applicability of Emotion Recognition and Induction Methods to Study the Behavior of Programmers. *Applied Sciences (Basel, Switzerland), 8*(3), 323. doi:10.3390/app8030323

Wrycza, S., & Maślankowski, J. (2020). Social Media Users' Opinions on Remote Work during the COVID-19 Pandemic. Thematic and Sentiment Analysis. *Information Systems Management, 37*(4), 288–297. doi:10.1080/10580530.2020.1820631

Xiao, H., Li, W., Zeng, G., Wu, Y., Xue, J., Zhang, J., Li, C., & Guo, G. (2022, January). On-Road Driver Emotion Recognition Using Facial Expression. *Applied Sciences (Basel, Switzerland), 12*(2), 807. doi:10.3390/app12020807

Yang, T., Wolff, F., & Papachristou, C. (2018). Connected Car Networking. *NAECON 2018 - IEEE National Aerospace and Electronics Conference,* 60-64. 10.1109/NAECON.2018.8556715

Yang, L., Tang, K., Yang, J., & Li, L.-J. (2016). Dense Captioning with Joint Inference and Visual Context. *Proceedings of the IEEE Conference on Computer Vision and Pattern Recognition (CVPR),* 1978–1987.

Yuksel, D., McKee, G. B., & Perrin, P. B. (2021). Sleeping when the world locks down: Correlates of sleep health during the COVID-19 pandemic across 59 countries. *Sleep Health, 7*(2), 134–142. doi:10.1016/j.sleh.2020.12.008

Zahra, Alivi, & Muazzam. (2022). Exam Anxiety among University Students, Jouranal of Management Practices. *Humanaties and Social Science, 6*(4), 19–29.

Zepf, S., Hernandez, J., Schmitt, A., Minker, W., & Picard, R. W. (2021, May). Driver Emotion Recognition for Intelligent Vehicles. *ACM Computing Surveys, 53*(3), 1–30. doi:10.1145/3388790

Zhang, J., Ling, C., & Li, S. (2019). *EMG Signals based Human Action Recognition via Deep Belief Networks.* Elsevier Ltd.

Zhang, C., Yu, M. C., & Marin, S. (2021, June). Exploring public sentiment on enforced remote work during COVID-19. *The Journal of Applied Psychology, 106*(6), 797–810. doi:10.1037/apl0000933

Zhang, X., Wang, R., Sharma, A., & Gopal, G. (2021). Artificial intelligence in cognitive psychology—Influence of literature based on artificial intelligence on children's mental disorders. *Aggression and Violent Behavior,* 101590. doi:10.1016/j.avb.2021.101590

Zubovich, N. (2022). *Advantages of Internet of Things: 10 Benefits You Should Know.* https://sumatosoft.com/blog/advantages-of-internet-of-things-10-benefits-you-should-know

About the Contributors

S. Geetha (Senior Member, IEEE) is currently a Professor and the Associate Dean, Research with the School of Computer Science and Engineering, Vellore Institute of Technology, Chennai Campus, India. She received the B.E. degree in Computer Science and Engineering from Madurai Kamaraj University, India, in 2000, and the M.E. degree in Computer Science and Engineering and the Ph.D. degree from Anna University, Chennai, India, in 2004 and 2011, respectively. Her research interests include steganography, steganalysis, multimedia security, intrusion detection systems, machine learning paradigms, computer vision and information forensics. She has more than 20 years of rich teaching and research experience. She has published more than 100 papers in reputed international conferences and refereed journals like IEEE, Springer, Elsevier publishers. She joins the Review Committee and the Editorial Advisory Board of reputed journals. She has given many expert lectures, keynote addresses at international and national conferences. She has organized many workshops, conferences, and FDPs. She was a recipient of the University Rank and Academic Topper Award in her B.E. and M.E. degrees, in 2000 and 2004, respectively. She was also the proud recipient of the ASDF Best Academic Researcher Award 2013, the ASDF Best Professor Award 2014, the Research Award in 2016, and the High Performer Award 2016, Best Women Researcher Award 2021.

D. Karthika Renuka is Professor in Department of Information Technology in PSG College of Technology. Her professional career of 18 years has been with PSG College of Technology since 2004. She is an Associate Dean (Students Welfare) and convenor for Students Welfare Committee in PSG College of Technology. She is a recipient of Indo-U.S. Fellowship for Women in STEMM (WISTEMM)-Women Overseas Fellowship program supported by the Department of Science and Technology (DST), Govt. of India and implemented by the Indo-U.S. Science & Technology Forum (IUSSTF). She was a Postdoctoral Research Fellow from Wright State University, Ohio, USA. Her area of specializations includes Data Mining, Evolutionary Algorithms, Soft Computing, Machine Learning and Deep Learning, Affective

Computing, Computer Vision. She has Organized an International Conference on Innovations in Computing Techniques Jan 22- 24, 2015 (ICICT2015) and National Conference on "Information Processing and Remote Computing" 27th and 28th February 2014 (NCIPRC 2014). Reviewer for Computers and Electrical Engineering, Elsevier, Wiley Book chapter, Springer Book Chapters on "Knowledge Computing and its Applications". She is currently guiding 8 research scholars for their Ph.D under Anna University. She has published several papers in reputed National and International journals and conferences.

Asnath Victy Phamila holds M.E and Ph.D degree in Computer Science and Engineering from Anna University, India. Her research area includes Image Processing, Computer Vision, Deep Learning, Wireless Sensor Networks and Network Security. She has around 17 years of academic and 3 years of industry experience. She has published around 40 research papers and she also serves as reviewer in reputed journals.

Karthikeyan N. holds M.E and Ph.D degree in Computer Science and Engineering from Anna University Chennai, India. His research area includes Distributed computing, Medical Informatics and IoT. He is having 20 years of experience in academic. He has published 15 research papers in reputed journals. 6 scholars are doing research under his supervision.

* * *

Arivarasi A. A. has more than 13 years of industry experience along with eight years of academic research experience. Her areas of expertise are sensors, IoT, 3D printing, and Machine learning/AI including digital transformation areas. She received her B.E. degree (in 2000) from Vellore Engineering College affiliated to Madras University, Chennai, India, and her Master of Engineering degree from Anna University, Chennai. She pursued her Ph.D. degree from BITS PILANI DUBAI Campus, UAE in 2019. Her Ph.D. thesis focussed on designing & developing an innovative process for 3D printing nanosensors for heavy metal sensors. She was the recipient of the Best Student Paper Award of International Conference on Electrical Engineering and Applications 2016, held at the University of California, Berkeley, US. She has published nine international journal articles on 3D printing research topics, presented research papers from ten international conferences on emerging technology-oriented topics, published a book chapter for World Scientific publisher on micro/nano 3D printing research, and is presently the reviewer for top journals - 'Rapid Prototyping' and 'Micro-Nano letters' IET, the journal of engineering.

L. Ashok Kumar is presently working as Professor in EEE Department at PSG College of Technology. He completed a Postdoctoral Research Fellow from San Diego State University, California. He is a recipient of BHAVAN fellowship from Indo US Science and Technology Forum. His current research focuses on Integration of Renewable Energy Systems in the Smart Grid and Wearable Electronics. He has 3 years of industrial experience and 17 years of academic and research experiences. He has authored 6 books in his areas of interest. He has published 110 technical papers in International and National Journals and presented 107 papers in National and International Conferences. He has completed 12 Government of India funded projects and currently 5 projects are under progress. His PhD work on Wearable Electronics bagged National Award from ISTE and he has received 12 Awards in the National level. He has 3 patents to his credit.

Bala Subramanian Chockalingam received his Bachelor of Engineering in Electronics and Communication Engineering from Anna University, Chennai by 2006. He received his Master of Engineering in Applied Electronics from Anna University, Chennai by 2008. He has completed Ph.D degree in the area of Wireless Sensor Network in the Department of Information Technology under Kalasalingam Academy of Research and Education by 2020. He is currently working as an Associate Professor in the Department of Computer Science and Engineering, Kalasalingam Academy of Research and Education. He has 14 years of experience in teaching and research. He has published more than 25 papers in reputed International Journals and Conferences. His areas of interest are Image and Signal Processing, Sensor Networks, Adhoc Networks, Internet of Things, etc.

Uma N. Dulhare is working as a Professor, Department of Computer Science & Artificial Intelligence, Muffakham Jah College of Engineering & Technology, Banjara Hills, Hyderabad. She has more than 20 years teaching experience. She has published more than 30 research papers in reputed National & International Journals & as a book chapter. She has received grants for 4 International Patents and published 2 National patents, One International copyright She is member of Editorial Board and reviewer of International Journals like IJACEA, ICDIWC and ICEOE, IIE, IJERTREW, IJDMKD, Elsevier Procedia. She is also the member of various professional societies like ISTE, CSTA of ACM, ASDF, IAENG & Fellow member of ISRD. She has also received a Best research paper Award in 2010, ASDF Global Award for Best Computer Science Faculty of the Year 2013 by the Lt. Governor of Pondicherry & also Best Academic Researcher of the year 2015. She honored with Outstanding Educator & Scholar Award 2016 by NEFD. Her area of interest is Networking, Database, Data Mining, Information Retrieval and Neural Networks & Big Data Analytics.

Nirmala G. is working as Assistant Professor in Department of Computer Science and Engineering in Kamaraj College of Engineering and Technology. She completed her UG degree in Computer Science and Engineering in 2005. She completed her PG degree in Computer Science and Engineering through Thiagarajar College of Engineering, Madurai in 2011. She completed her Doctoral degree in Information and Communication Engineering through Anna University, Chennai in 2021. Her area of interest includes Machine Learning and Artificial Intelligence. She has 15 years of experience in teaching profession. She has published more than 25 papers in International Journals and Conferences.

Niranjana G., Professor, Department of CSE in SRM Institute of Science and Technology, has finished B.E from Madras University, M.Tech (CSE) from SRM University and Ph.D in Computer Science and Engineering from SRM University. She is having nearly 20 years of teaching experience. She has published around 20 Scopus Indexed papers. She has received Young Investigator Award in 2012, Best Paper Award in 2016 and IET Women Engineer Award in 2018. Her area of interest includes Networking, Machine learning,Deep Learning and image processing.

Sandhia G. K. is currently working as an Assistant Professor in SRMIST, Chennai. She obtained her PhD in SRMIST and published papers in reputed journals. The author is specialized in the area of network security, machine learning, Cloud Computing. The author has many papers on cryptography which is one of the important area of specialization.

Akshay Giridhar is a final (fourth) year undergraduate student of Vellore Institute of Technology, Chennai. His specialization is Electronics and Communication Engineering. His major research domains are Data Analytics, Machine Learning, and Blockchain.

Alagiri Govindasamy is the director for two start-up companies, Future Connect Technologies, and PMCGS Private Ltd., both are focusing on Industry 4.0 consulting, product development, and competency building initiatives for corporates. He has secured 25.6 years of corporate experience in turnaround consulting, large-scale ERP implementation, and Digital transformation initiatives for signature clients across the globe. He has completed his MS from BITS Pilani - India, MBA from Manchester Business School – UK, and pursuing his Doctor of Business Administration (DBA) program from Nottingham Business School – UK. He is the visiting faculty to leading business schools like BITS, Pilani – India, NMIMS, Mumbai – India, SP Jain School of Global Management – Sydney/Singapore/Dubai locations.

Hemanth Harikrishnan is an undergraduate student at Vellore Institute Of Technology (VIT University) - Chennai Campus. He is currently pursuing his Bachelor of Technology degree which is focused in Electronics and Communication Engineering, and he is presently in his final year of study. His domain of interest includes Artificial Intelligence and Signal Processing, which he aspires to explore through vivid problem statements. His research interests include Data Analytics, Machine Learning, and Signal processing. More specifically his interest is towards insights driven approach for signal processing using AI. He has earlier presented his work in the domain of Machine Learning.

Hemalatha J. received her BE and ME in Computer Science and Engineering under Anna University Chennai in 2007 and 2013. She is currently pursuing her PhD in Information and Communication Engineering from Anna University, Chennai, India. In July 2013, she joined the Department of Computer Science and Engineering at P.S.R. Engineering College, Sivakasi. India. Currently she is an Assistant Professor in Department of Computer Sceince and Engineering, Kalasalingam Academy of Research and Education, India. She has published more papers in reputed international conferences and journals. Also she is the author of the book chapter: Combating Security Breaches and Criminal Activity in the Digital Sphere, A Volume in the Advances in Digital Crime, Forensics, and Cyber Terrorism (ADCFCT) Book Series by IGI Global. And also published six book chapters. Her research interests include digital steganography, steganalysis, machine learning and image processing.

Nimala K. is working as Assistant Professor in Department of Networking and Communications in School of computing, SRM Institute of Science and Technology, Kattankulathur, Chennai, India. She obtained her Doctorate in Computer Science Engineering from SRMIST, India in 2020. Her research interest include Machine Learning, Deep Learning, Cloud Computing, Analytics etc. Her main focus of research include Sentiment Analysis, Topic modelling, Predictive analytics and big data."

Jansi K. R. is working in SRM Institute of Science and Technology, Kattankulathur, Tamil Nadu, India.

Umashankar Kumaravelan received his B.Tech in Computer Science and Engineering from Vellore Institute of Technology in 2021. He is currently planning to pursue his higher education in the field of Artificial Intelligence. His main interests lie in the field of computer vision.

Nivedita M. is currently pursuing Ph.D in Computer Science and Engineering, VIT-Chennai Campus, Chennai in the area of Computer Vision and Image Processing. She has completed her M.E. from Anna University, Chennai. She has published 4 papers in international journals. Her areas of interest are Image Processing, Computer Vision, and Artificial Intelligence.

Uma Maheswari received her M.E in Computer Science and Engineering from the Madras University, Chennai, India in 2002 and Ph.D. in Information and Communication Engineering in 2011 from Anna University, Chennai. Currently, she is working as a Professor in the Department of Computer Science and Engineering at the P.S.N.A. College of Engineering and Technology, Dindigul, India. She has totally 16 years of teaching experience which includes 12 years of research experience. Her research interests include Biometrics, Image processing, Compiler design, Artificial Intelligence, Speech Processing, and Wireless Sensor Networks. She has published 50 papers in International journals, 2 papers in National journals, and presented 30 papers at International conferences, and 20 papers at National conferences. She has co-authored a book entitled "Compiler Design" and "Theory of Computation" published by Yes Dee Publishing. She is a recognized Ph.D. supervisor in Anna University of Technology in the area of Image processing, Cloud computing, Network security and Networks. Acted as a reviewer for various referred journals, acted as a Coordinator for various seminars, conferences, etc.

Sekar Mohan is a well-known academician has been taken as charge of principal AAA College of Engineering and Technology. He has obtained his doctoral degree from Kyungpook National University. He was awarded the prestigious Korean Research Fund to carry out his research. He has published more than 50 SCI papers and more Scopus indexed papers.

Ramkumar Narayanaswamy is pursuing a full-time Ph.D. at PSG College of Technology. He completed his M.E Computer Science Engineering at PSG College of Technology and Master of Science in the University of Leicester, United Kingdom. He has six years of experience as an Assistant professor and three years of Industrial experience. He has published 15 papers in international journals and conferences.

Jeya R. is working in Department of Computing Technology, School of Computing, SRMIST, Kattankulathur, Tamil Nadu, India.

Muthuselvi R. is working as Professor in Department of Computer Science and Engineering in Kamaraj College of Engineering and Technology. She completed her UG degree in Electrical and Electronics Engineering through Thiagarajar College of Engineering in Madurai in 1989. She completed her PG degree in Computer Science and Engineering through MEPCO SCHLENK College of Engineering, Sivakasi in the year 2001. She completed her Doctoral degree in Information and Communication Engineering through Anna University, Chennai in 2012. Her area of interest includes High Performance Computing and Internet of Things. She has more than 25 years of experience in teaching profession. She has published more than 30 papers in International Journals and Conferences.

Nareshkumar R. was born in 1987. He received her B.E. degree in Computer Science and Engineering from Anna University, India, and M.E degree in Computer Science Engineering from Anna University, India. He has 11 years of experience in teaching. Currently, He is currently pursuing the PhD degree in Computer Science and engineering with the School of Computing, SRM institute of science and technology, Chennai, Tamilnadu, India. His research interests include affective computing, natural language processing, deep learning, machine learning, and human–machine interaction.

Vidhya R. is currently working as an Assistant Professor in SRMIST, Chennai. She obtained her PhD in SRMIST and published papers in reputed journals. The author is specialized in the area of machine learning and learning analytics.

Vidhyapriya R., Professor & Head, Biomedical Engineering Department, PSG College of Technology, received her B.E. degree in Electrical and Electronics Engineering and Masters in Applied Electronics from PSG College of Technology, Coimbatore. She is awarded Ph.D from Anna University Chennai for her research in the field of Wireless Sensor Networks. With 21 years of teaching experience, she is also proactively engaged in research in emerging technologies and has successfully completed sponsored research projects funded by DRDO, AICTE and UGC. The outcomes of the research have contributed to patents and research publications. She has to her credit three patents and has successfully converted the prototype of her research work into products through technology transfer. She has published around 60 papers in national and International Journals. She has organized International Conferences, several National Level Conferences, Faculty development Programmes and Workshops. She is also the executive council member of the PSG Tech Alumni Association.

Christhu Raj is working in M. Kumarasamy College of Engineering, Karur, Tamilnadu, IN.

Shaik Rasool received Master of Technology in Computer Science & Engineering from Jawaharlal Nehru Technological University in 2011. He is currently working as Assistant Professor in the Department of CSE at Methodist Collage of Engineering & Technology, Hyderabad, Telangana, India. He has more than 10 Years of Teaching Experience. He has received grants for 4 International Patents and published 2 National patents, One International copyright, 7 Book Chapter in International Publications, One research paper in International Conference and 12 research papers in various International Journals. He is member of various National and International Professional Bodies. He completed 4 R & D Projects funded by MJCET, Hyderabad, India. He received Best Senior Faculty Award in National Faculty Award 2021-2022 for academic contribution in Computer Science and Engineering by Novel Research Academy, Puducherry, India on 14 December 2021. His research areas of interest include Artificial Intelligence, Machine Learning, Data Science, Internet of Things, Cloud Computing, Big Data, Data Mining, and Information Security.

Shanthalakshmi Revathy J., has received her B.E in Information Technology from Mohamed Sathak Engineering College and received M.E in Computer Science and Engineering from PSG College of Technology, Coimbatore in 2010. She is currently working as Assistant Professor in the Department of Computer Science and Engineering at Velammal College of Engineering and Technology, Madurai, India. She has more than 15 publications and 2 book chapter, 1 book in National, International Conference, and peer-reviewed International Journal proceedings. Her research interests span a broad range of interesting topics, includes machine learning, deep learning, predictive analytics and evolutionary computing.

Nagadevi S. is working in Department of Computing Technology, School of Computing, SRMIST, Kattankulathur, Tamil Nadu, India.

Arun Sahayadhas is an Professor with the Department of Computer Science and Engineering in Vels Institute of Science, Technology and Advanced Studies, Chennai, India. He has completed his undergraduate and post graduate degrees from Annamalai University, Chidambaram and his doctorate degree from University Malaysia Perlis also known as UniMAP in Perlis, Malaysia. During his research days, he was a part of the 'Artificial Intelligence – Rehabilitation (AI-Rehab) Research Group' and has worked and guided research projects in different areas of Artificial Intelligence. His contribution in terms of publications in indexed conferences and

JCR impact factor journals, reflect his passion towards research. He is a reviewer in many impact factor journals and has also served as an Editor in a Scopus indexed journal proceeding. He has a DST research project worth Rs 30 Lakhs. He has filed one patent. He also serves as the Director of Internal Quality Assuarance Cell at the Institution and thereby enables quality culture.

S. Sasikala is currently working as an Associate Professor in Department of Computer Science and Engineering, Velammal College of Engineering and Technology Madurai, TamilNadu, India. She received doctorate in faculty of Information and Communication Engineering from Anna University India, 2016. She has published more than 25 Journal and Conference papers in the area of Data mining and Big Data Analytics with Elsevier Science Direct, Springer and IEEE publishers. She has published two International Scientific books in KDD and Data mining and Data warehousing. She is serving as an Editorial Board Member and Reviewer for many reputed journals like IEEE, ELSEVIER and SPRINGER. She has 19+ years experience in teaching and Research. Her research interests include Data Mining, Internet of Things, Machine Learning Paradigms and Optimizations.

Sathiya Narayanan Sekar has more than 10 years of academic research experience. His areas of expertise are signal processing, computer vision, machine learning and data analytics. He received his B.E. degree (in 2008) from Anna University, Chennai, India, and his M.S. and Ph.D. degrees (in 2011 and 2017 respectively) from Nanyang Technological University (NTU), Singapore. His Ph.D. thesis focussed on developing novel compressive sensing reconstruction algorithms for sparse signals with partially known information. He was the recipient of the NTU Research Student Scholarship 2011-2015 and the NTU-EEE Outstanding Teaching Assistant Award 2014. From October 2015 to April 2018, he served as a Research Associate/Fellow in NTU. The project was on compressive sensing applications to radar imagery and video, and it was funded by Singapore's Ministry of Education. Since June 2018, he is working as an Assistant Professor (Senior Grade) in Vellore Institute of Technology, Chennai.

Jerritta Selvaraj received her BE, ME and PhD degrees in Electronics from Manonmaniam Sundaranar University, Anna University, and Universiti Malaysia Perlis (UniMAP) respectively. Currently she works as the Professor and Head of the Department of Electronics and Communication engineering, Vels Institute of Science, Technology and Advanced Studies (VISTAS), Chennai, India. Her research areas include intelligent signal processing, non-linear and data dependent signal processing and analysis, machine learning etc., that focuses on children and persons with special needs. She is a member of the Artificial Intelligence Research Lab at VISTAS and works on various projects.

Arjun Sharma is a final (fourth) year undergraduate student of Vellore Institute of Technology, Chennai. His specialization is Electronics and Communication Engineering. His major research domains are Machine Learning, Natural Language Processing and Behaviour Analytics.

Utkarsh Singh is an undergraduate student at Vellore Institute of Technology (VIT University) - Chennai Campus. He is currently pursuing his Bachelor of Technology degree which is focused in Electronics and Communication Engineering, and he is presently in his final year of study. His domain of interest includes Data Science and Blockchain, which he aspires to explore professionally through vivid corporate projects via imbibing knowledge regarding relevant tech-stacks or in general, skillsets. His research interests involve Data Analytics, Machine Learning, and Blockchain. More specifically, his research interests involve incorporating a data-driven approach in interpreting stuffs and attaining a standpoint which is technically feasible and robust from business point of view. He has earlier presented a research work in the domain of blockchain.

Rajeev Sukumaran, Engineering Epistemologist, is a 'Learning Researcher'. Having a field experience of 30+ years in academia and industry, he had been instrumental in consulting several large-scale innovative educational and technology projects. In Education; he contributes to: Curricula Development, Instructional Design Systems and Technologies, Scientific Teaching-Learning, Technology-Enabled Learning, Educational Leadership. He helps Educational Institutions & Industry in analyzing, designing, developing, implementing and evaluating educational and skill programmes towards developing learning and inculcating specific learning and behavioral skills. In Industry; he contributes towards: establishing learning culture, learning competency framework, scientific recruitment, leadership succession, rapid technology learning acumen, and research-productivity mindset. Currently is the Director, Learning and Development at SRM Institute of Science and Technology, India.

G. Suseela received the B.E., M.Tech., and Ph.D. degrees in computer science and engineering. She is currently working as an Assistant Professor with the School of Computing, SRM Institute of Science and Technology, Chennai, Tamil Nadu, India. She has 14 years of academic experience and four years of research experience. Her research interests include wireless image sensor networks, image-processing, deep learning, and network security. To her credit she has around 25 publications in various international journals.

Om Prakash Swain is a final (fourth) year undergraduate student of Vellore Institute of Technology, Chennai. His specialization is Electronics and Communication Engineering. His major research domains are Artificial Intelligence, Data Analytics and Computer Networking.

Nithyananthan V. is currently working as Assistant Professor in Directorate of Learning & Development, SRM Institute of Science and Technology, Tamil Nadu, India. Graduated UG in B.A. Corporate Secretaryship, PG in M.Com., MBA, M.Sc. Yoga, awarded Ph.D. in Management. Interested research area on Emotional Intelligence, Work Life Balance, and Quality of Life Satisfaction. He has more than two decades of experience in the field of HR Management, Spiritual education, conducted workshop and session more than 1200 on yoga, meditation, Spiritual education to the school and college students, teachers, employees especially Tamil Nadu Govt. service around 18,000 employee, and 46,000 from village people through World Community Service Centre.

Vivek V. completed his B.Tech (IT) degree from National College of Engineering, Tirunelveli in the year of 2011 and Completed his M.E., (CSE) from Manonmaniam Sundaranar University, Tirunelveli in the year of 2014. Also completed his Doctor of Philosophy (Ph.D.) in Computer Science and Engineering under Manonmaniam Sundaranaar University- Tirunelveli in the year of 2018. Currently he is working as Associate Professor in AAA College of Engineering and Technology. He has published more than 13 papers in various reputed International Journals. He is an author of many Scopus indexed book chapters as well.

Index

Recommended Reference Books

IGI Global's reference books can now be purchased from three unique pricing formats:
Print Only, E-Book Only, or Print + E-Book.
Shipping fees may apply.

www.igi-global.com

ISBN: 9781799834991
EISBN: 9781799835011
© 2021; 335 pp.
List Price: US$ 225

ISBN: 9781799836612
EISBN: 9781799836636
© 2021; 287 pp.
List Price: US$ 215

ISBN: 9781799841654
EISBN: 9781799841661
© 2021; 354 pp.
List Price: US$ 225

ISBN: 9781799851011
EISBN: 9781799851028
© 2021; 392 pp.
List Price: US$ 225

ISBN: 9781799871569
EISBN: 9781799871583
© 2021; 393 pp.
List Price: US$ 345

ISBN: 9781799870784
EISBN: 9781799870807
© 2021; 368 pp.
List Price: US$ 195

Do you want to stay current on the latest research trends, product announcements, news, and special offers?
Join IGI Global's mailing list to receive customized recommendations, exclusive discounts, and more.
Sign up at: **www.igi-global.com/newsletters.**

Publisher of Timely, Peer-Reviewed Inclusive Research Since 1988

www.igi-global.com　　Sign up at www.igi-global.com/newsletters　　f facebook.com/igiglobal　　t twitter.com/igiglobal

Ensure Quality Research is Introduced to the Academic Community

Become an Evaluator for IGI Global Authored Book Projects

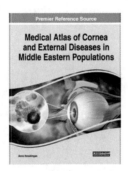
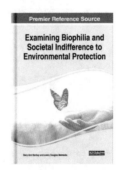
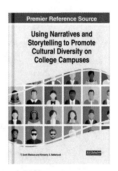

The overall success of an authored book project is dependent on quality and timely manuscript evaluations.

Applications and Inquiries may be sent to:
development@igi-global.com

Applicants must have a doctorate (or equivalent degree) as well as publishing, research, and reviewing experience. Authored Book Evaluators are appointed for one-year terms and are expected to complete at least three evaluations per term. Upon successful completion of this term, evaluators can be considered for an additional term.

If you have a colleague that may be interested in this opportunity, we encourage you to share this information with them.

Easily Identify, Acquire, and Utilize Published
Peer-Reviewed Findings in Support of Your Current Research

IGI Global OnDemand

Purchase Individual IGI Global OnDemand Book Chapters and Journal Articles

For More Information:

www.igi-global.com/e-resources/ondemand/

Browse through 150,000+ Articles and Chapters!

Find specific research related to your current studies and projects that have been
contributed by international researchers from prestigious institutions, including:

- Accurate and Advanced Search

- Affordably Acquire Research

- Instantly Access Your Content

- Benefit from the InfoSci Platform Features

" *It really provides* **an excellent entry into the research literature of the
field**. *It presents a manageable number of* **highly relevant sources** *on topics of
interest to a wide range of researchers. The sources are* **scholarly, but also
accessible** *to 'practitioners'.* "

- Ms. Lisa Stimatz, MLS, University of North Carolina at Chapel Hill, USA

Interested in Additional Savings?

Subscribe to

IGI Global OnDemand *Plus*

Learn More

*Acquire content from over 128,000+ research-focused book chapters and 33,000+ scholarly journal
articles for as low as US$ 5 per article/chapter (original retail price for an article/chapter: US$ 37.50).*

6,600+ E-BOOKS.
ADVANCED RESEARCH.
INCLUSIVE & ACCESSIBLE.

IGI Global e-Book Collection

- **Flexible Purchasing Options** (Perpetual, Subscription, EBA, etc.)
- Multi-Year Agreements with **No Price Increases** Guaranteed
- **No Additional Charge** for Multi-User Licensing
- No Maintenance, Hosting, or Archiving Fees
- Transformative **Open Access Options** Available

Request More Information, or Recommend the IGI Global e-Book Collection to Your Institution's Librarian

Among Titles Included in the IGI Global e-Book Collection

Research Anthology on Racial Equity, Identity, and Privilege (3 Vols.)
EISBN: 9781668445082
Price: US$ 895

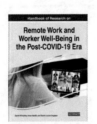

Handbook of Research on Remote Work and Worker Well-Being in the Post-COVID-19 Era
EISBN: 9781799867562
Price: US$ 265

Research Anthology on Big Data Analytics, Architectures, and Applications (4 Vols.)
EISBN: 9781668436639
Price: US$ 1,950

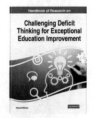

Handbook of Research on Challenging Deficit Thinking for Exceptional Education Improvement
EISBN: 9781799888628
Price: US$ 265

Acquire & Open

When your library acquires an IGI Global e-Book and/or e-Journal Collection, your faculty's published work will be considered for immediate conversion to Open Access *(CC BY License)*, at no additional cost to the library or its faculty *(cost only applies to the e-Collection content being acquired)*, through our popular **Transformative Open Access (Read & Publish) Initiative**.

For More Information or to Request a Free Trial, Contact IGI Global's e-Collections Team: eresources@igi-global.com | 1-866-342-6657 ext. 100 | 717-533-8845 ext. 100

Have Your Work Published and Freely Accessible

Open Access Publishing

With the industry shifting from the more traditional publication models to an open access (OA) publication model, publishers are finding that OA publishing has many benefits that are awarded to authors and editors of published work.

Freely Share
Your Research

Higher Discoverability
& Citation Impact

Rigorous & Expedited
Publishing Process

Increased
Advancement &
Collaboration

Acquire & Open

When your library acquires an IGI Global e-Book and/or e-Journal Collection, your faculty's published work will be considered for immediate conversion to Open Access *(CC BY License)*, at no additional cost to the library or its faculty *(cost only applies to the e-Collection content being acquired)*, through our popular **Transformative Open Access (Read & Publish) Initiative**.

Provide Up To
100%
OA APC or
CPC Funding

Funding to
Convert or
Start a Journal to
**Platinum
OA**

Support for
Funding an
**OA
Reference
Book**

IGI Global publications are found in a number of prestigious indices, including Web of Science™, Scopus®, Compendex, and PsycINFO®. The selection criteria is very strict and to ensure that journals and books are accepted into the major indexes, IGI Global closely monitors publications against the criteria that the indexes provide to publishers.

WEB OF SCIENCE™    Compendex   Scopus®

PsycINFO®    Inspec

**Learn More
Here:**

For Questions, Contact IGI Global's Open Access
Team at openaccessadmin@igi-global.com

Are You Ready to
Publish Your Research **?**

IGI Global
PUBLISHER of TIMELY KNOWLEDGE

IGI Global offers book authorship and editorship opportunities across 11 subject areas, including business, computer science, education, science and engineering, social sciences, and more!

Benefits of Publishing with IGI Global:

- Free one-on-one editorial and promotional support.

- Expedited publishing timelines that can take your book from start to finish in less than one (1) year.

- Choose from a variety of formats, including Edited and Authored References, Handbooks of Research, Encyclopedias, and Research Insights.

- Utilize IGI Global's eEditorial Discovery® submission system in support of conducting the submission and double-blind peer review process.

- IGI Global maintains a strict adherence to ethical practices due in part to our full membership with the Committee on Publication Ethics (COPE).

- Indexing potential in prestigious indices such as Scopus®, Web of Science™, PsycINFO®, and ERIC – Education Resources Information Center.

- Ability to connect your ORCID iD to your IGI Global publications.

- Earn honorariums and royalties on your full book publications as well as complimentary copies and exclusive discounts.

Join Your Colleagues from Prestigious Institutions, Including:

Australian National University

Massachusetts Institute of Technology

JOHNS HOPKINS UNIVERSITY

HARVARD UNIVERSITY

TSINGHUA UNIVERSITY 1911

COLUMBIA UNIVERSITY
IN THE CITY OF NEW YORK

Learn More at: www.igi-global.com/publish

or by Contacting the Acquisitions Department at: acquisition@igi-global.com

Printed in the United States
by Baker & Taylor Publisher Services